W9-AFN-677

EPIC TOMATOES

EPIC TOMATOES

HOW TO SELECT & GROW *the* BEST VARIETIES *of* ALL TIME

CRAIG LeHOULLIER

Storey Publishing

The mission of Storey Publishing is to serve our customers by publishing practical information that encourages personal independence in harmony with the environment.

EDITED BY **Carleen Madigan**
ART DIRECTION AND BOOK DESIGN BY **Carolyn Eckert**
TEXT PRODUCTION BY **Liseann Karandisecky**
INDEXED BY **Christine R. Lindemer, Boston Road Communications**

COVER PHOTOGRAPHY BY **© GAP Photos, Ltd./Amanda Darcy, front and spine; Mars Vilaubi, end papers (hardcover edition); © Shoe Heel Factory, back (author) and inside front; © Stephen L. Garrett, back (all except author) and inside back**
HAND-LETTERING ILLUSTRATIONS BY **© Mary Kate McDevitt, except Carolyn Eckert, 35, 36, 40, 108, 109, 115**

LETTERS AND EPHEMERA COURTESY OF THE AUTHOR
INTERIOR PHOTO CREDITS APPEAR ON PAGE 255

Storey Publishing
210 MASS MoCA Way
North Adams, MA 01247
www.storey.com

Printed in Canada by Transcontinental Printing
10 9 8 7 6 5 4 3 2 1

Library of Congress Cataloging-in-Publication Data

LeHoullier, Craig.
Epic tomatoes / by Craig LeHoullier.
pages cm
Includes bibliographical references and index.
ISBN 978-1-61212-208-3 (pbk. : alk. paper)
ISBN 978-1-61212-464-3 (hbk. : alk. paper)
ISBN 978-1-61212-209-0 (ebook) 1. Tomatoes. I. Title.
SB349.L44 2015
635'.642—dc23

2014029010

DEDICATION

To my wife and best friend, Susan — for her unwavering love, support, and encouragement.

To my dear daughters, Sara and Caitlin — now growing their own tomatoes and on their own life adventures. I am so proud of you all.

contents

PREFACE

This book has been a long time coming. Tomatoes, or at least gardens, have played a significant role in my life for many years. One of my first recollections is of carrying a watering can around in a neighbor's driveway at the tender age of two. I spent many weekend days at a local park riding on my father's shoulders as he lifted me up to smell the tree blossoms. At other times, he walked me through the flower beds, pointing out the vividly blooming dahlias, zinnias, and marigolds.

My grandfather Walter, an avid and creative gardener, delighted me with tours of his backyard plot, which was full of sweet peas, strawberries, and, of course, tomatoes. In fact, the first I sampled were the succulent, red fruits of his garden, bursting with flavor. Little did I know it at the time, but that revelation would set me on a lifelong journey, one that will last as long as I can plant a seedling.

A GARDEN OBSESSION GROWS

Life intervened between those very early years in the gardens of my loved ones and the first of my own creation. Yet, once again, the association of gardening with family, love, and wonder arose. The spring following our marriage, Susan and I rented a nearby community plot set aside for graduate students and embarked on creating our first garden. Filled with flowers, herbs, and vegetables, tended to as our young daughter Sara slept in her stroller, it was a great success. Most of the plants we grew were from starts purchased at the local garden center. Hence our vegetables were not at all uncommon — Bell Boy peppers and Better Boy tomatoes, for example. But we grew them ourselves, they tasted wonderful, and we learned a lot through the experience.

Through the years, we gardened in New Hampshire, Seattle, and suburban Philadelphia, in addition to our current location in Raleigh. We perused the varieties in local gardening centers and learned each year a few things that helped and other things that didn't. However, pretty quickly I started to grow bored with the fruits of our labors. It was clear, even early on, that people garden for many reasons. Some aim for the biggest vegetables, some the earliest, and others for the highest yielding or the tastiest, no matter what the specific type of crop. I realized within the first few years of gardening that uniqueness, variety, historical relevance, and flavor are the most important criteria for me. Perhaps it was my tendency to take a scientific approach to my hobbies, treating them as projects, but, clearly, purchasing a six-pack of Better Boy tomatoes or a packet of green Blue Lake bush beans just wouldn't suffice.

Little did I know that gardening was already becoming something of an obsession. After deciding that the nursery center offerings left me wanting, my attention turned to seed catalogs, which meant learning how to successfully start all sorts of seeds. At that time, I was also a subscriber to a few good gardening magazines, and they pointed to a wonderfully diverse selection of seed companies to try to greatly expand my gardening experience. It turned out that the mid-1980s was a perfect time to catch the serious gardening bug. Along with the standby companies such as Burpee and Parks, smaller, more experimental companies were popping into view, such as Johnny's, Seeds Blum, and Gleckler. The range of colors, sizes, shapes, and flavors grew exponentially if you were willing and able to start your own seedlings. But even when considering seeds from seed companies, constraints remained in terms of variety. Most companies focused on hybrid varieties, and all companies had a need to be profitable and therefore ensure that their selections were commercially viable.

ENTER THE HEIRLOOMS

At about the same time that I was making the switch from nursery-purchased transplants to self-started seedlings, I read a magazine article about the Seed Savers Exchange (SSE), which enticed me to join. That significant decision revolutionized my gardening efforts. Each of my major criteria — historical significance, uniqueness, variety, and flavor — could be satisfied through my involvement with the SSE. I joined in 1986, and today the relevance of the organization and the impact on what we grow has never been clearer.

Switching from mostly hybrids to mostly heirlooms felt a bit risky. If one were to take the words of many of the catalogs as written, anyone who didn't grow mostly or exclusively hybrid varieties was doomed to gardens of diseased or dead plants with disappointing yields. I sensed an interesting challenge and opportunity for some research, so I set about growing some of the most highly regarded hybrid tomatoes alongside an even larger selection of non-hybrid and heirloom varieties that sounded attractive to me.

Burpee's Matchless
Mexico Midget
Bisignano #2
Lillian's Red Kansas Paste
Favorite
Arkansas Traveler
Ferris Wheel
Mortgage Lifter
Polish
Black Krim
Cherokee Chocolate
Japanese Trifele Black
Egg Yolk
Lillian's Yellow Heirloom
Hugh's
Tiny Tim Yellow
Sun Gold
Cherokee Green
Tiger Tom
Red Zebra
Green Giant
Ruby Gold
German Striped

Big Boy
Nepal
Giant Syrian
Magnus
Stump of the World
German Johnson
Rosella Purple
Dwarf Wild Fred
Lemon Drop
Golden Queen
Sweet Sue
Summertime Gold
Blazing Beauty
Yellow Oxheart
Yellow Brandywine
Kellogg's Breakfast
Green Zebra
Speckled Roman
Don's Double Delight
Lucky Cross
Coyote
Yellow White

Running over three gardening seasons, the results I obtained clearly showed that the "must grow the hybrid" contention was just not true. (See page 228.) The non-hybrids/heirlooms I grew equaled or out-yielded the hybrids in general, with far superior flavors and variety. And that came with the benefit of being able to save seed for next year's garden.

It was my good fortune to have joined SSE in 1986, just as it was getting its feet firmly planted into the soil. My early focus on tomatoes meant that among the 1,000-plus varieties I've grown throughout the years, the vast majority were shared by other, similarly enthusiastic tomato growers in transactions made through the SSE member network. I read the annual yearbook, picked out those varieties that particularly caught my interest, obtained the seeds, grew them (often in several locations as we moved about), and then saved and, in turn, shared my own seeds. I retained each and every letter that accompanied the seeds, and some of those original letters are excerpted in this book (pages 240–244) to highlight what I consider are especially important tomatoes. My understanding of tomato colors, flavors, shapes, and sizes grew through experience and hands-on growing of each of the varieties that are described in this book. As my interests evolved from growing out the best of the historically relevant heirlooms, through happening upon some unique finds in my own garden, to finally co-conceiving and leading a project to create new, very useful dwarf-growing tomato varieties (see page 47), my immersion in the world of tomatoes has only deepened with time.

Old American seed catalogs provide a valuable glimpse into the early days of tomato development. This page demonstrates the tendency to exaggerate that we still see today.

This book is a distillation of what I've learned and experienced throughout 35 years of gardening. I will consider it to be a success if it excites, educates, and inspires readers to grow tomatoes themselves. Whether in a large garden or in small containers on a deck or driveway, there is simply no excuse to not grow your own to demonstrate what a culinary marvel a tomato can be. I have a warning, though: You may find yourself living without fresh tomatoes between your last harvest of the autumn and the first of the following spring. You will realize that those round, pinkish-red things that are available in the stores bear no resemblance at all to a real tomato, for no other type of homegrown produce so far exceeds in quality that which is purchased.

My hope is that, in reading this book, you will be inspired to embark on a gardening journey that will renew you each season and last a lifetime. You'll come away from each garden anxious to begin the following season, with many successes and a few inevitable misfortunes on which to build and learn. Each year, my own gardening efforts teach me something new, reinforce the validity of tried-and-true practices, and bust a few urban gardening myths. I want to share what I've learned so that everyone who has the desire to grow and taste a tomato that is delicious or colorful or interesting can do so. Pick up a shovel, grab a pot, put on the garden gloves, and let's grow some great tomatoes.

Ten Tastiest

NEPAL

Though nothing special to look at, **NEPAL** is the tomato that converted me from hybrids to heirlooms. It is a tomato that is perhaps most like those special tomatoes tasted in our youth, obtained at local farm stands or a relative's garden. It has an aggressive, full flavor that will delight those who love intense tomatoes.

INTRODUCTION

We who are gardening these days find ourselves in a unique position in agricultural history. We have not only the best developments of the plant breeders in the forms of hybrids of high quality, adaptability, and disease tolerance, but an incomparable selection of historical varieties that have been preserved through the efforts of countless gardeners. Our grandparents and great-grandparents would be envious of the number of different varieties from which we can choose today. The range of tomato colors, shapes, sizes, and flavors is astounding. Their foliage may be pale green to nearly sickly yellow, medium forest green, or saturated to a point where it looks nearly blue. The plants may grow as tall as a house or be as short as the span of a child's hand. The leaves are toothed or smooth, ridged or crinkly. And since seed saving from tomatoes is so simple, the seeds can be shared with others, allowing the passing on of living history.

Tomatoes are my ultimate horticultural focus. From pea sized to humongous, from round to heart-shaped, and from squat to frying pepper–like, with colors that seem to be infinite in possibilities, diversity in all respects rules the day. Upon cutting one open, you might observe a dominance of meat or of seeds. You can see whether it's dense or juice-laden, thick skinned or delicate, ugly or stunning, faulty or faultless. Most of the attributes reside in the genes, and the weather, season, and cultural practices add their own character. Tasting a variety of different homegrown tomatoes is much like tasting a range of distinctly different beers, wines, or chocolates. There are dominant flavor notes and a whole host of nuances and subtleties. The sampling of a tomato becomes a thought-provoking exercise in adjective seeking. The intensity, balance, tart/sweet flavors, fruitiness, sharpness or flatness. The right tomato can move you to tears, or to search for new adjectives, or to look for a place in which to spit out the remains. Different tomatoes conjure up all kinds of cooking creativity, working backward to formulate a recipe around the tomato itself rather than using the tomato as a flavor enhancer.

In addition to all that is the rich legacy associated with many of the best, most interesting varieties. When you walk out into your garden you can imagine the stories of the various people and families who made growing these wonderful varieties possible. The garden can be a supplier of food, but it can also be a laboratory for experiments and tests, and a museum of living history. All of it is really quite wonderful, and this only begins to express the joy that I find in growing tomatoes. It is a pursuit that renews itself each season, and I am delighted to have you along on the journey.

AT RIGHT: **These American seed catalogs from the early 1900s depict tomatoes that are still in existence today, thanks to the efforts of seed savers.**

Livingston's Seed
Annual
1900
Our new
Tomato
for 1900
Livingston's
Magnus.
"True
Blue"
Seeds.
The Livingston Seed Company,
Columbus, Ohio, U.S.A.

38th Year
Burpee's
Annual
1914
The
Plain Truth
About
Seeds
That Grow.
Burpee's
"Matchless" Tomato
for 25 years
the best main-crop tomato.
Registered
WAB
Trade Mark
"The Seal of Quality."
Seed Farms in
Pennsylvania,
New Jersey
and California.
Copyright 1913 by
W. Atlee Burpee & Co.,
Burpee Buildings,
Philadelphia.

Tomato Growing
on
H. W. Buckbee's
Rockford Seed Farms
Rockford, Ill., U.S.A.
Buckbee's
Beefsteak
Tomato
Delicious
Solid Meat.
Matures early,
hardy grower,
a splendid shipper,
enormously productive,
best of all,
large size,
handsome color,
the best keeper.
Buckbee's
Beefsteak
Tomato
Pkt. $ .10.
3 Pkts. .25.
½ oz. .35.
1 oz. .60.
2 oz. 1.10.
¼ lb. 2.00.

Established 1850
Livingston's
Seed Annual
1921
Livingston's
Globe
Pkt. 10¢
True Blue
Seeds
Livingstons
Stone
Pkt. 10¢
The
Livingston
Seed Co.
"Famous
for
Tomatoes"
Livingston's
Golden Queen
Pkt 10¢
114
North High St.
Columbus
Ohio

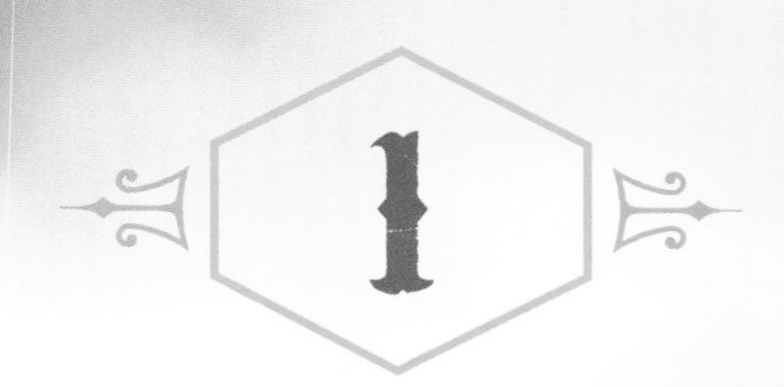

The Origins of TODAY'S TOMATO

IMAGINE KITCHEN GARDENS at different points in time. If we look back 5,000 years, we might find corn, potatoes, peppers, and onions. (Indeed, we could take it back another 5,000 years and still find potatoes.) What is missing, of course, is the tomato. Few people today are likely to realize the relative youth of tomatoes in the gardening and culinary world. Here is how it happened.

TOMATOES COME TO NORTH AMERICA

The Mayans and other Mesoamerican people domesticated the tomato plant and first used it in cooking. A mutation was likely responsible for converting the small two-chambered wild types into the larger, lumpy, multi-chambered fruit that represents the vast majority of today's tomatoes. It is the large tomatoes that were nurtured and developed by Central American farmers. The Aztecs named the plant *xitomatl,* or large *tomatl.*

How it was introduced to British North America is a matter of speculation; there are a number of possibilities. The Spanish (who brought tomatoes back from Mexico during the 1520s and then distributed them throughout the Spanish empire and into Asia) consumed tomatoes; they'd had settlements in Florida earlier in the seventeenth century, which could account for tomato introduction into what is now Georgia and the Carolinas. French Huguenot refugees and British colonists could have also brought seeds directly to the Carolinas. Or tomatoes could have come here from the Caribbean, since migration from the British West Indies to the southern colonies began in the late seventeenth century. Some historians consider the Caribbean route to be the most likely, because the term *tomato* was part of the local language. Whatever the mode of introduction, tomatoes were cultivated in the Carolinas by the middle of the eighteenth century.

"THAT SOUR TRASH"

By the early nineteenth century, tomatoes were present in many towns across America, though it appears most people didn't eat them, for a variety of reasons. First, many colonists from England, Scotland, and Ireland came here before the tomato was commonly eaten in their homeland. Next, America was physically isolated from the rest of the world leading up to that time, so most people had no idea how popular the tomato was becoming in Europe. Finally, even where there was expanding knowledge of how to grow tomatoes, few recipes were published in books prior to the 1820s.

Poison?

The tomato had a reputation for being poisonous (and some believe this to be true even today) but actual published opinions of such are very rare prior to 1860. In fact, Andrew Smith, in doing research for his excellent book *The Tomato in America*, found only a few such references. Among them was a reprinted British medical work that represented obsolete beliefs in England and a statement made by Thomas Jefferson's grandson indicating that the tomato was considered poisonous when his grandfather was young. It's interesting to note that the number of references to the potential toxicity of tomatoes rose significantly after 1860.

Perhaps stories of the poisonous nature of tomatoes arose from people's attempts at drama and intrigue and speculation, rather than facts. In 1852, a physician named Neil Lewis made a presentation in Cincinnati in which he criticized tomatoes. After the talk, a number of people came forward to offer their own complaints. Some testified to suffering from a peculiar condition of the stomach, piles, and tender, bleeding gums and loose teeth, all from eating tomatoes.

One story was of a young woman who claimed to have lost all of her teeth after eating a quantity of tomatoes.

The most famous early American tomato story is the daring public tomato consumption — perhaps — by Robert Gibbon Johnson in Salem, New Jersey, reported to have occurred in 1820. Johnson, one of Salem's prominent citizens, grew tomatoes in his garden from seeds obtained from South America. He planned to publically consume tomatoes that he himself grew, which of course (since tomatoes were widely considered poisonous) created quite a sensation. Hundreds of onlookers reportedly traveled from far and wide to witness this remarkable event. The story goes that Johnson bit into a tomato, some onlookers fainted, and, with Johnson suffering no ill effects, the tomato industry in America began.

Offending the Senses

Why else did many people shun tomatoes in America prior to the popularity boom of the late 1800s? Some did not like the smell of the tomato plant itself. One woman in upstate New York considered them a beautiful plant but would have preferred cooking "ripe potato balls" than eating the tomatoes themselves, thinking that the odor of the plant was sufficient warning against consumption of the fruit.

Others just didn't like how they looked. It's truly remarkable how the mere sight of a ripe tomato offended so many gardeners. In the 1820s, one gardener in Massachusetts is recorded to have said, "The first time I saw a tomato, they appeared so disgusting that I thought I must be very hungry before I am induced to taste them."

In the late 1820s in Pennsylvania, J. D. Garber noted that no more than two people in a hundred, on first tasting, would ever be persuaded to taste "that sour trash" a second time. The dislike of the tomato was not limited to northerners. A North Carolinian, Charles Blietz, in 1831, described a dinner at an inn outside of Richmond, Virginia. He was served sliced red tomatoes (his first experience eating them), and he notes how it spoiled his dinner. He wrote that statement in 1887, confirming that his dislike of tomatoes did not fade with time. S. D. Wilcox of Florida, a noted agriculturist, ate his first tomato in 1836 as the major component of a pie. He said that it was an "errant humbug and deserves forthwith to be consigned to the tomb of the Capulets."

However, there are many historical instances of tomatoes, either the fruit or extracts and concoctions, being promoted as an agent of health. Because of this, it is surmised that many of those who grew fond of tomatoes started with them in such a manner.

It is also abundantly clear that the flavor of the tomato is often an acquired taste. (I'm sure many of us remember not liking tomatoes when we were children.) The editor of *The American Farmer,* an early nineteenth-century jounal, thought that most people found the flavor disagreeable at first (primarily because of the acidic flavor sensation) but grew to enjoy tomatoes over time. Even Scottish-born John Muir, soon to become one of our preeminent naturalists, noted that English and Scottish settlers in Wisconsin in the 1850s had nothing but contempt for tomatoes, which were "so fine to look at with their sunny colors, but so disappointing to the taste." Ralph Waldo Emerson, in 1856, agreed that the taste for tomatoes was an acquired one.

DON'T EAT YOUR VEGGIES

The shunning of tomatoes actually extended to the fear of eating vegetables in general. Historian Richard Hooker thought that this was because of a frequent practice of washing the garden's bounty in polluted water, or because the high time for vegetable consumption, summertime, coincided with various illness epidemics, in which case the vegetables became guilty by happenstance.

MEET THE VARIETIES
TOMATOES WITH WILD ORIGINS

Sometimes great things come in small packages. Both of these unique tomatoes originated in Mexico, likely have wild genes in their heritage, and came to me as wonderful surprises in 1990. Both have become real favorites with many of my seedling customers.

Mexico Midget

Barney Laman is a gardener from Chico, California, whom I got to know through some *National Gardening* magazine seed swaps starting in the late 1980s. In 1990, I received a letter from Barney with a packet of seeds, which he claimed were those of the world's smallest tomato. The seeds were collected by Barney's brother, a New Mexico trucker, on one of his hay-delivering trips to Texas, from someone who told him that they originated in Mexico. Mexico Midget, which is the name of the variety, is as tiny as a pea. It grows like a weed and volunteers freely, though, oddly, it is not at all easy to germinate on purpose. Hidden in the tiny red round fruit are all of the intense, rich, wonderful flavors of the largest, most treasured beefsteak-type variety.

Coyote

The story of the tomato Coyote is very different. During an heirloom tomato demonstration at the Pennsylvania Horticultural Society Harvest Show of 1990, I was approached by a sweet woman named Maye Clement holding a complete cluster of tiny, nearly white cherry tomatoes still attached to the vine. She freely offered me the unique-looking variety, saying that it was from Mexico. I gratefully accepted the tomato gift and attended to saving seed as soon as I could get home.

The following year, I grew the tomato for the first time and was overwhelmed with the vigor, productivity, and beauty of the variety. The flavor is quite unusual, with sweetness predominating, along with an unusual component that some people love and others are less fond of. My desire for more information on the tomato was answered a year later, when I received a letter from Ms. Clement. She related that the tomato hails from and grows wild in Veracruz, Mexico, where it is called Coyote or Tomatito Silvestre Amarillo (wild small yellow tomato). Like Mexico Midget, it volunteers readily, and once gardeners grow it, they have it for as long as they wish to carry on with the various babies that show up in following seasons.

Coyote (which I erroneously listed as Ivory Currant in the SSE yearbook prior to receiving the true name) looks like no other tomato. Though not quite pea sized, it is still smaller than most cherry tomatoes. The rampant vines produce clusters of nearly translucent fruit that are ivory at the blossom end, shading to pale yellow at the shoulders. The flavor is certainly an acquired taste for many, but we like it just fine. It is beautiful and versatile, and it carries with it a nice story — how can you ask for more in a tomato?

TOMATOES GAIN ACCEPTANCE

A remarkable thing happened in America when, starting in the 1820s, the tomato found its way into more and more gardens, recipes, and restaurants. An article published by Dr. John Bennett in 1834 extolling — with numerous distortions and exaggerations — the health benefits of tomatoes seemed to mark the beginning of the wave of tomato popularity that continues, unabated, today. Certainly, tomatoes were an acquired taste, and once they were accepted, and especially supported by various health claims, it is no wonder that the call for more different and improved varieties arose to meet the ever-expanding recipes and uses.

TOMATO VARIETIES OF EARLY AMERICA

Judging from how it was portrayed in the paintings of the day (see Art Imitates Life), the tomato of early America was closer in shape and form to many of the treasured heirloom types grown today than most probably realize. What we are less certain about is the range of color and flavor, since the relative novelty of the tomato meant that there was nothing to which to compare those very first varieties available in the mid-1800s.

Tomatoes were sold, with primary demands coming from French immigrants, by the Landreths in Philadelphia in the 1790s, and the first tomato seeds were being sold by 1800. By the 1830s, tomato seeds were sold throughout the country. Looking at old seed catalogs reveals that the "varieties" were often distinguished by no more than a color and size. Efforts to improve tomatoes began with a need for early ripening, in order to avoid the significant price drop that would occur when tomatoes would arrive en masse at the peak of the season and glut the market.

Early Seed Saving

Early efforts at improving tomatoes focused on saving seed from particular fruits that exhibited

ART IMITATES LIFE

Perhaps the best way to tell what early tomatoes looked like is to examine illustrations in old research papers, catalogs, and even paintings.

1544 – **The first European reference to a tomato appears. Italian herbalist Pietro A. Matthioli notes that "mala aurea" (golden apple) was a flattened fruit, shaped like an apple, and ripened from green to gold. A red variety was later described in 1554.**

1553 – **A painting by Gesner and a Dodoens woodcut portray a large, lumpy red fruit.**

1826 – **A painting called *Still Life with Vegetables* by James Peale shows a large tomato the size of an apple with lobes, lumps, and creases. A still life from 1864 by Paul LaCroix shows tomatoes of similarly lumpy character.**

Still Life with Vegetables **by James Peale, circa 1826**

A HISTORY OF CULTIVATED TOMATOES

In the Beginning

The tomato is assumed to have originated in the coastal highlands of western South America. Wild versions — very small and with only two seed cavities — still exist in the coastal mountains of Peru, Ecuador, and northern Chile.

1520s

The Spanish encounter the tomato during the Cortez conquests of Mexico. Once back in Spain, the tomato is distributed throughout the empire, where it makes its way to Asia.

1529

One of the first written examples of the culinary use of tomatoes was by a Jesuit priest who lived in Mexico. He notes the tomato as being "very wholesome, full of juice, and good for sauce or eating." Also in 1529, another priest in Mexico said that the Aztecs combined tomatoes with chiles and ground squash seeds to make a sauce, or salsa.

1597

John Gerard's *Herball* first documents tomatoes being grown in England, planted by Gerard in his gardens in Holborn. It notes that "golden apples or apples of love" came from Spain and Italy, and "while the leaves are toxic, the fruit is not."

1608

The first European culinary use of tomatoes is recorded, as a component of a salad with cucumbers.

1692

The earliest known cookbook with tomatoes included is published in Naples, the recipes being of Spanish origin.

1753

Linnaeus (Carl von Linné) places the tomato as part of the family Solanaceae, with the scientific name *Solanum lycopersicon.* Linnaeus's friend Philip Miller renamed it *Lycopersicon esculentum* (edible wolf peach) in 1754.

Early 1700s

Tomatoes are widely consumed in Jamaica and throughout the Caribbean.

1710

English herbalist William Salmon records the first known reference to the tomato in the British North American colonies. In his book *Botanologia*, he notes that he observed tomatoes growing in "Carolina," which was actually the southeast part of Florida.

1771

At this point, the word *tomato* has emerged (probably originating in Jamaica, before it was adopted in England and British North America).

Early seed catalogs went to fanciful extremes to try to convince customers that their tomato varieties were the most appealing. Sadly, several of the varieties on this page appear to be extinct.

a desired characteristic. This proved to be unsuccessful, and it wasn't until Alexander Livingston's breakthrough — using single plant, rather than single fruit, selection — that a reliable tomato-improvement method arose. Another method of tomato improvement resulted in a true breakthrough. Dr. T. J. Hand of New York crossed cherry tomatoes with the large, lumpy varieties that were common at the time. It is assumed that Dr. Hand then spent some time growing out his crosses and carrying out selections to achieve his particular goal, which ended up being the variety Trophy, a tomato with a "solid mass of flesh and juice, small seeds, and smooth skin." The promoter, Colonel George Waring, sold seeds of Trophy for 25 cents apiece.

Others became involved in selecting and breeding tomatoes, and some of these early efforts are documented in Fearing Burr's *Field and Garden Vegetables of America,* initially published in 1863. In the book, Mr. Burr lists 24 tomatoes varieties. Even at this early date, it is clear that there was confusion about tomato names, as well as color. Some descriptions offered multiple names for certain varieties, or indications that a particular tomato appeared to be essentially the same as another. Simply from reading the text it is difficult to confirm whether the tomato is red (yellow skin over red flesh) or pink (clear skin over red flesh). It is comforting — or frustrating — to note that the same ambiguous discussions of tomato varieties that are commonplace today have been going on for 150 years!

Naming Early Tomatoes

Many of the early tomato names are fairly unimaginative and pretty much descriptive of the appearance. It was many years before the advent of names such as Big Boy that indicated the supposedly superlative nature of a given seed company's newest creations.

Here are some of the tomatoes listed in Burr's 1863 guide, with synonyms in parentheses:

Apple Tomato (Apple Shaped), Bermuda, Cook's Favorite, Fejee, Fig (Red Pear), Giant (Mammoth), Grape (Cluster), Improved Apple, Large Red, Large Red Oval, Large Yellow, Lester's Perfected, Mexican, Red Cherry, Red Plum, Round Red, Round Yellow, Seedless, Tree Tomato (de Laye), White, White's Extra Early (Early Red), Yellow Cherry, Yellow Pear Shaped, and Yellow Plum.

The Vegetable Garden of 1885 by MM. Vilmorin-Andrieux listed 22 varieties of tomato:

Apple Shaped, Apple Shaped Purple (Acme), Apple Shaped Red (Hathaway's Excelsior), Belle de Leuville, Cherry, Early Dwarf Red, Favorite, Green Gage, Jaune Petite, King Humbert, Large Red, Large Early Red, Large Yellow, Pear Shaped (Fig), Red Currant, Rose Colored Smooth Criterion, Smooth Red Curled Leaf, Stamford, Tree (de Laye), Trophy, Turk's Cap, and Yellow Pear.

Alexander Livingston Refines the Selection Process

How did we get from the paltry selection of tomatoes available as recently as the 1880s to the stunning and overwhelming collection of tomatoes that we are able to select from today, the number of which is likely over 10,000 in terms of unique names? The seed houses at the time did their best to improve the tomato, but it was the work of Alexander Livingston that provided the breakthrough.

Misconceptions abounded when considering efforts to improve the tomato prior to 1870. Most felt that by simply saving seeds from a particular tomato that exhibited a distinct characteristic — be it earliness, smoothness, or lack of cracking — and growing them the following year, the tomatoes on the plants that resulted would move toward the improvement. As an example, if a particular variety ripened its fruit later than one hoped, saving seed from the first ripe tomatoes on the plants would, with any luck, result in an earlier variety. The problem with the theory of single-fruit selections is that all tomatoes from a particular plant contain seeds that are essentially the same genetically. And this is why tomato improvements were so stubborn, until the efforts of Mr. Alexander Livingston of Reynoldsburg, Ohio.

Livingston, the founder of what would become the Livingston Seed Company, had a revelation: rather than focusing on the fruits, he looked at large plantings of tomatoes and selected a particular plant that showed a favorable characteristic. He correctly assumed that there was something genetically different about that particular plant, and that if seeds were saved from tomatoes from that plant, an improved variety was possible, given some further selection work. In truth, a different plant in a fairly uniform planting could well be a rare mutation, or the result of cross pollination; either way, it was an appropriate starting point for new varieties.

Using his new single-plant selection technique, Livingston went on to revolutionize the tomato in America. Between his first new tomato introduction, Paragon (1870), and the later Globe (1906), the Livingston Seed Company introduced a set of tomatoes that represented significant improvements on what had come before, typically in terms of relative smoothness and uniformity of shape. This was at a time when many tomatoes were irregularly shaped, leading to considerable waste in the canning process (during a period when canning was a burgeoning industry and very important to many farmers and home gardeners for ensuring tomato supplies during the off season).

The Livingston Seed Company created many of the most widely grown, important new tomato varieties between 1879 and 1930. Globe, Beauty, and Stone are still available today and are indeed very good tomatoes.

MEET THE VARIETIES

KEY LIVINGSTON INTRODUCTIONS

These three wonderful tomatoes are true American heirlooms that originated from Alexander Livingston, the world's most creative tomato-breeding mind in the late 1800s. My tomato-growing friends and I have helped bring them back into existence by utilizing the tomato-seed collection in the Germplasm Resources Information Network (GRIN) database at the USDA. The tomato seeds in this collection represent an opportunity to bring historic varieties back into circulation. Growing these wonderful old tomatoes, and sharing their seeds with other gardeners, is the most important step in ensuring that they live on forever.

Favorite

Livingston released Favorite in 1883 with the goal of satisfying the growing needs of the tomato canning business. Most early American tomato varieties tended to be irregular in shape and size — not ideal for canning. Livingston selected Favorite as a single plant growing in a field of Paragon, another of his company's varieties.

According to Livingston, Favorite was an early, blood-red, smooth, and prolific tomato that was uniformly solid (having no air spaces in the seed cavities), ripened evenly, and had firmer, thicker flesh than other tomatoes of that time. Testimonials quoted by Livingston in his catalogs and book show that Favorite was immediately a very popular variety.

I found Livingston's Favorite listed in the tomato collection in the GRIN database and obtained a sample of seeds to grow in 1994. Favorite was indeed a delightful surprise, and the tomatoes the plants produced matched the descriptions in the Livingston seed catalogs. What I find interesting is the desire for fairly ordinary-looking tomatoes of medium size and red coloring, because of the need for tomato preservation through canning. The irony is that the very tomatoes heirloom enthusiasts now seek out (large, unusually colored, often irregularly shaped) are just the types that were not favorable in the late 1800s and early 1900s. Changing times mean changing tastes and changing tomato uses.

Magnus

Magnus is quite a special tomato. Aside from allowing us to take a peek at real horticultural history, the combination of pink fruit and potato-leaf foliage makes it quite unusual. The indeterminate plant produces a fine yield of nearly round, blemish-free, half-pound fruit that possess a lovely sweetness and excellent balance.

On top of that, this great Livingston introduction tells another sort of story — a more personal one, and a testament to friendship, curiosity, perseverance, and luck.

While at a conference in New Hampshire one year, I came across a wonderful antiques store, where I purchased a pristine seed catalog from the Livingston Seed Company, dated 1900. The company, specializing in tomatoes, released new varieties most every year and depicted them on their catalog covers. The 1900 Livingston cover tomato was a variety called Magnus. From the picture, it was clear that the tomato was pink (quite unusual for this era) and had potato-leaf foliage (extremely rare).

Purchasing that catalog launched my quest to find Magnus. Unfortunately, it was a fruitless pursuit for many years. Magnus appeared to be one of those tomato varieties that had simply become extinct. I had begun to turn my attention to finding other lost tomatoes when I first heard of the USDA tomato seed collection and the ability to search it using the GRIN database. After many search attempts, I was finally able to uncover a treasure trove of the original Livingston varieties — the seeds of which had been donated to the collection by seed companies in the 1940s — including Favorite, Perfection, Paragon, Golden Queen, Dwarf Stone, and, yes . . . Magnus!

Upon receiving my request for seeds the USDA sent me an alarming quantity (nearly 100 seeds). Apparently, their collection was growing old and seed germination rates would be low. My tomato-growing friend Carolyn Male and I tried our hand at growing Magnus, but it was clear that the seeds were indeed knocking on death's door. My efforts resulted in one seedling, and it showed regular leaf foliage; clearly, it was seed from a crossed fruit, or stray seed that was mixed in.

Carolyn also obtained one seedling, but it showed the proper potato-leaf foliage. Generously, Carolyn carefully packed and shipped the plant to me, and Magnus found a place in my garden. The plant grew well, and the fruit ripened to provide lovely, nearly round, medium-size pink specimens that possessed a really fine flavor, well balanced but slightly on the sweet side.

I saved plenty of seeds, listed Magnus in the SSE yearbook, and sent samples to a few seed companies. Magnus was nearly lost once, and I was determined that our stroke of luck would not be for nothing.

Golden Queen

Golden Queen was the first widely available, well-regarded yellow tomato released to the public by a seed company. Livingston developed the variety from seed he acquired at a country fair from a tomato simply described as a "very pretty yellow tomato."

Prior to the release of Golden Queen in 1882, few yellow tomatoes were listed in seed catalogs, and those offered were likely a result of crosses and thus not stable (few were in catalogs for more than a few years). This means that we will never know which fruit Livingston started with when he developed Golden Queen.

Interestingly, several other old tomatoes I obtained from the USDA — such as Golden Monarch and Golden Beauty — grew out to be very similar to Golden Queen. You could imagine that other seed companies, jealous of Livingston's yellow tomato, grew out Golden Queen themselves and made their own selections for release under slightly different names.

Seed catalogs in the 1980s often listed a variety called Golden Queen but described it as a determinate tomato, or one that was more of an orange hue. Golden Queen is certainly a fine name for a tomato, and it is likely that the original Golden Queen simply fell by the wayside at modern seed companies, and its name was reused for a very different tomato.

The seeds of the original Golden Queen, which I acquired from the USDA and grew out, produced tall, vigorous plants with oblate, medium-size, bright yellow fruits. Occasional larger fruits possess a faint pink blush, which is noted in some of the Livingston catalog descriptions. And, though I think that in calling it the "best-flavored tomato in existence" Livingston was laying on the praise a bit thick, it is certainly delicious, and among the best flavored of the true bright yellow varieties we have at our disposal.

Ten Tastiest

YELLOW OXHEART

Sometimes seed catalog superlatives translate to real experiences. I purchased seeds for this **YELLOW OXHEART** because of the praise heaped on it when it was first listed in the Southern Exposure Seed Exchange catalog. Not only beautiful to behold, with its large, truly heart-shaped, rich yellow fruit, it delights the senses with dense, succulent flesh and a perfectly balanced flavor.

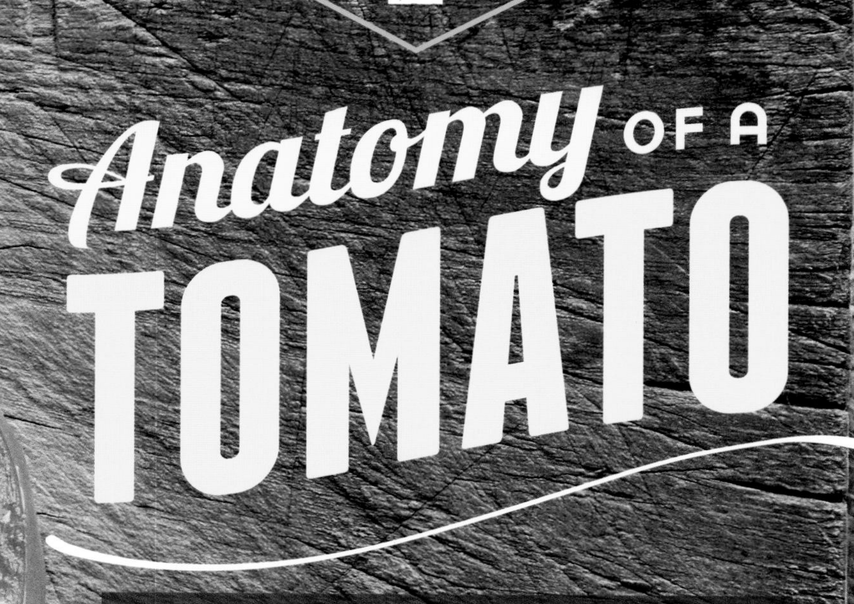

2 Anatomy of a TOMATO

BRITISH JOURNALIST Miles Kington once said, "Knowledge is knowing that a tomato is a fruit. Wisdom is not putting it in a fruit salad." But as clever as that sounds, the lines are now blurred, as we contemplate tomatoes so sweet that they can be made into ice cream or festoon our dinner salads, with strawberries and pears sitting happily alongside the Sun Golds and Mexico Midgets and wedges of Tiger Toms. Knowing exactly what to call tomatoes is not important. What is important is having the ability, through gardening or shopping, to make the most out of such a time as this, when the variety of tomatoes available is essentially infinite. Our journey begins here.

SHAPE, SIZE, AND USE

Tomatoes range from perfectly round (globe shaped) to highly flattened (referred to as oblate) — as if someone put one hand on the blossom end and another on the stem end and squeezed to flatten the fruit. They can be oval or egg-shaped (also called "deep globe"), long and slender like a frying pepper, heart-shaped, or even pear-shaped, with a thin neck and bulging bottom. Many of the larger oblate tomatoes have irregularities, creasing, and an overall individuality of form (which to some can veer toward ugliness) that many people associate with the classic shapes of some of the older, heirloom tomatoes.

From Pea to Grapefruit Size

The round tomatoes range from the size of a pea to that of a softball or grapefruit, and from a fraction of an ounce to 2 or 3 pounds. The flattened, oblate types range from 1 or 2 ounces to 3 pounds or more. Egg-shaped tomatoes start at an ounce and can reach half a pound (larger in rare cases), as do the longer, thinner specimens. Heart-shaped tomatoes range from half a pound to 2 pounds or more. Of course, there are varieties of tomatoes of sizes throughout the extremes noted here. The realm of tomato options seems to grow unabated year to year, as more heirlooms surface and amateur breeders work to fill in any gaps.

A Tomato for Every Use

Various terms are used to describe the preferred uses of tomatoes by shape. Tiny or small round tomatoes are often called cocktail, grape, cluster, or cherry tomatoes. Such tomatoes are wonderful for salads, omelets, frittatas, kebabs, or pizza toppings. Tomatoes of a few ounces are sometimes known as salad, breakfast, or grilling tomatoes, and any tomatoes that work well on a sandwich — whether half a pound or multi-pound monsters — are often called slicing tomatoes.

Since the late 1800s, the medium-size round tomatoes seem to have been preferred for canning. Prior to that date, though, very large, irregular varieties that originated in Italy and found their way to California were the canning choice. The meaty, elongated, slender tomatoes often consisting of a dense flesh with less juice are favored for sauces or pastes; hence the term "paste tomato." Finally, many of the large tomatoes, when sliced parallel to the blossom and stem ends, show a dense, meaty flesh with small seed-filled cavities. For years, these have been known as beefsteak tomatoes, a term first used in the late 1800s. The designation has nothing to do with flavor, and everything to do with comparison to the density of a slice of steak.

Each type of tomato — whether juicy, dry, soft, succulent, firm, or meaty — interacts with tasters' palates in different ways, and one person's cherished variety could be another's to toss into the compost heap. This is the fun of tomato tastings. For the cook, part of the fun of creation during harvest time is to match a type of dish with what's ripe that day.

TOMATO COLORS

The diversity in tomato shapes and sizes is fascinating, but the real fun arrives when considering the incredible rainbow of color options available in tomatoes today. I enthusiastically tasted my first non-red tomatoes in 1987, when I grew Persimmon (pale orange), Czech's Excellent Yellow and Lemon Boy (bright yellow), Ruby Gold and Pineapple (yellow with red swirls), Tiger Tom (red with gold zigzag vertical stripes), and Brandywine (pink). Waiting for the fruit to ripen was so exciting, and the beauty of the fruit inspired us to find ways to take advantage of their appearances in our cooking exploits.

What Color *Is* That?

Though much of the confusion surrounding tomato colors comes from inexperience, even the most experienced gardeners struggle with

factors such as the unclear terminology, personal opinions on what a particular color name connotes, and perhaps just a difference in the ways that some of us perceive and identify color. It doesn't help that tomato color classification errors are very common in seed catalogs and on gardening websites. Confusion over classification of tomato varieties by color is not recent, and seed catalogs from the very beginning don't appear to have used standard or consistent terminology.

Another important point about color is that a tomato's outside appearance is actually the combined effect between skin color and flesh color. In fact, skin color is critical in determining how we classify a tomato, with a single exception (see Green and White, page 37). The net effect is that we call a tomato a particular color given what it looks like unsliced. A cut slice of the tomato would often be called something quite different, and there would be a more limited color palette if we categorized tomatoes by their flesh, rather than the color of the uncut fruit.

Finally, many will wonder what the relationship is between tomato color and flavor. It is up to each gardener and tomato taster to reach his or her own conclusions, but to my taste buds, the flavor of each specific tomato variety is distinct (sometimes subtle, sometimes screamingly so), and though some generalizations can be made, so many exceptions exist that it is better to not use color as a guide to tomato flavor.

Red tomatoes represent most people's idea of tomato color and are therefore the yardstick of comparison for those who have yet to be exposed to the fancifully colored heirloom types. The defining characteristic of red tomatoes is yellow skin color. The effect of yellow skin stretched over red flesh makes the tomatoes appear to be scarlet, the hue of red with just a trace of orange. The flesh of red tomatoes can vary widely, ranging from a rather pale pink to a deep crimson. The combination of yellow skin over red flesh represents dominant gene expression, which explains the relatively large number of red tomato varieties. Red tomatoes typically predominate at larger farmers' markets, because the red hybrid types such as Celebrity and Mountain Pride are often preferred for field planting and large-scale production.

REPRESENTATIVE RED TOMATOES: Big Boy, Better Boy, Roma, Celebrity, and Big Beef (hybrids); Andrew Rahart's Jumbo Red, Aker's West Virginia, and Abraham Lincoln (heirlooms).

If red flesh is covered with clear, rather than yellow, skin, the tomato is referred to as pink (although crimson could be an even more descriptive designation). There are comparatively fewer pink tomatoes because the clear-skinned characteristic is a recessive trait. The pink/red designation tends to be quite confusing for many gardeners, possibly because of the varying abilities to define colors or just inexperience with seeing the differences between the two.

In old seed catalogs, pink tomatoes were often called "purple," because the types we know of today as closer to true purple (such as Cherokee Purple) weren't known at the time. With no good comparison, to some early eyes the pink tomatoes, when compared to red ones, looked purple. This can also be seen in the realm of many pink-fruited heirloom tomatoes, such as Pruden's Purple, which are clearly misnamed as to color.

Red and pink tomatoes can be distinguished only by their skin. The color of the flesh of each type is quite similar.

REPRESENTATIVE PINK TOMATOES: German Johnson, Pink Brandywine, Ponderosa, Eva Purple Ball, and Anna Russian.

One recent tomato phenomenon is the significant increase in the number of black tomatoes, starting in the mid-1990s. What distinguishes black varieties is very deep crimson-red flesh combined with varying proportions of retained green pigment, particularly in the gel surrounding the seeds. Black tomatoes possess a recessive genetic trait that calls for retention of some chlorophyll even after full ripening, resulting in a green-over-red effect. This leads to significantly darker coloration. If the skin color is clear, the apparent color of the tomato is purple, with a distinct darkening at the shoulders. If the skin color is yellow, the apparent color is a burnt mahogany/chocolate brown. Since experience with these two relatively recently occurring tomato colors is still somewhat limited, the error rate in color attribution is high; many purple tomatoes are classified as brown or chocolate, and vice versa.

As with red and pink tomatoes, once purple and brown tomatoes are sliced, their inner appearance is identical. If you wish to take advantage of the distinct difference in outward appearance, it is necessary to ensure that the tomato pieces are large enough and presented so that the skin is easily seen.

REPRESENTATIVE PURPLE TOMATOES: Cherokee Purple, Black Krim, Black from Tula, Purple Calabash, and Black Cherry.

REPRESENTATIVE BROWN TOMATOES: Black Plum, Black Prince, Cherokee Chocolate, and Japanese Trifele Black.

Red:
NEPAL

Pink:
FERRIS WHEEL

Brown:
CHEROKEE
CHOCOLATE

Purple:
BLACK KRIM

Yellow:

GOLDEN QUEEN

Orange:

SUN GOLD

Green:

GREEN GIANT

White:

YELLOW WHITE

A continuum of hue exists between the palest of yellow and deepest of orange tomatoes, and this is quite a tricky color category to deal with. Tomatoes with yellow flesh covered by clear skin can approach the bright color of a goldfinch or fade to a color closer to ivory. Varieties with yellow skin and yellow flesh range from a pale hue to a deep shade of butter yellow. The ripe fruit color of a particular yellow or orange variety can vary from season to season, or even within a particular season, depending on the temperature at which the tomato ripens. Throw in color terms such as "gold," and clear categorization becomes problematic. Different people call all sorts of colors "gold," from pale yellow to a rich near-orange yellow. When the flesh darkens to orange, a clear-skinned variety will become pale orange, similar to a pumpkin, and yellow-skinned, orange-fleshed varieties will appear as a deep, rich orange.

When served sliced, tomatoes in this color category can become quite interesting. Depending on skin color, some tomatoes that appear orange from the outside have bright yellow interiors because of the skin/flesh interaction. Some yellow tomatoes, when sliced, appear to be nearly white. It all makes for great fun when preparing a simple caprese salad in a way that places a rainbow of variety on the plate.

REPRESENTATIVE YELLOW TOMATOES: Yellow Pear, Lemon Boy, Lillian's Yellow Heirloom, and Lemmony.

REPRESENTATIVE ORANGE TOMATOES: Sun Gold (hybrid); Yellow Brandywine, Persimmon, and Kellogg's Breakfast (heirlooms).

Most unusual in the spectrum of tomato colors are those that exhibit such a pale yellow shade as to be nearly white or ivory, as well as the varieties that remain green when in edible ripe condition. The so-called white varieties have near-white flesh and a clear skin. They often develop a pale pink blush at the blossom end, and when they are very ripe, they take on more of a pale yellow color. This is one area where there does seem to be a relationship between tomato color and flavor. With very few exceptions, I've found the flavor of white tomatoes to be quite mild, verging on bland.

Green-fleshed tomatoes can have either yellow or clear skin. I've found green-fleshed tomatoes to be almost consistently quite wonderful and among my favorites for fresh eating.

The green-fleshed varieties indicate their ripeness by turning a rich orange-yellow color, which makes categorizing them somewhat confusing. Despite their yellow outward appearance as they hang on the plant, they are classified as green tomatoes because of their ripe *flesh* color.

The green-fleshed varieties with clear skin provide a true picking-time challenge for the gardener. But these varieties do typically develop a pale pink blossom-end blush and take on a subtle, hard-to-describe color change when ready for eating. The squeeze test works well, too.

REPRESENTATIVE WHITE TOMATOES: Great White, Snow White Cherry, Dr. Carolyn, White Beauty, and White Queen.

REPRESENTATIVE GREEN TOMATOES: Cherokee Green, Dorothy's Green, and Evergreen (yellow skin); Green Giant and Aunt Ruby's German Green (clear skin).

To complete the remarkable range of tomato colors, some varieties display more than one color when ripe. Tomatoes with yellow flesh swirled with pink or red, both inside and out, are often referred to as bicolored beefsteak types. On rare occasion, the swirled color combination is green with deep crimson, appearing as nearly purple with green shading externally, typified by the variety Ananas Noire. This class of tomatoes is easy to spot even when sliced, because the mix of colors bleeds into the flesh, as well as being apparent externally.

MEET THE VARIETIES

LILLIAN'S YELLOW HEIRLOOM

In 1988, I received three small packets of seed, along with a brief but pleasant letter, from Robert Richardson of New York. One of the seed packets was labeled "Lillian's Yellow Heirloom #1" and contained a local heirloom variety that Lillian Bruce, a gardener from Tennessee, had grown for many years.

Lillian's Yellow Heirloom

Lillian's Yellow Heirloom is a very vigorous, indeterminate plant with enormous, dark bluish-green, potato-leaf foliage. The fruit are smooth, oblate, and very large (averaging 12 ounces, but often over 1 pound). They ripen very late (105 days from transplant) but are well worth the wait. The color is a very pale yellow with just the faintest hint of pearly pink at the bottom of the tomato (which was especially unique for potato-leaf varieties when I first grew it), and the flavor is outstanding. My experience in 1989 was certainly very limited compared with that of today, but just about all yellow or orange varieties I knew then tended to be very mild, sweet, or bland. That is not at all the case with Lillian's Yellow Heirloom. I'd wager that a blindfolded tomato taster would conclude that the variety is one of the best he or she had ever tasted, and would likely guess it to be a red or pink variety— certainly not bright yellow.

One surprise, not at all welcomed by a seed saver, is the variety's dearth of seeds. A typical 1-pound fruit of Lillian's Yellow Heirloom will likely provide less than 25 seeds. Given the high ratio of flesh to gel and seeds, one might think the texture would be too firm and dry, but nothing is further from the truth: few tomatoes are as succulent and enjoyable to eat. Each year my wife and I eagerly anticipate the days when some slices of Lillian's Yellow Heirloom are available to feature in our evening meals.

After my initial experience with Lillian's Yellow Heirloom, I listed it in the SSE yearbook. Eventually it ended up in a few seed catalogs, and it has gained in popularity over the years. I've also been delighted to discover that it really enjoys our challenging North Carolina weather and does consistently well in our gardens each year.

Lillian's Red Kansas Paste

When Robert Richardson sent me the seed of Lillian's Yellow Heirloom, he also included a variety that he called Lillian's Red Kansas Paste. All we know about the variety is that Lillian Bruce received the seeds some years previously from someone in Kansas and subsequently shared them with Mr. Richardson. Lillian Bruce considered it the best-tasting tomato in her collection.

This is a tomato that deserves wider recognition. Produced on a tall, vigorous plant with the typical wispy foliage and floppy stems of most paste and heart-shaped varieties, the brilliant scarlet red tomatoes are irregular in shape, tending to oval, but not quite heart-shaped and not quite paste. And unlike typical paste-type tomatoes, they have plenty of juice, lots of seeds, and a full-bodied, rich flavor that makes them a delight in salads and a star in sauces.

Finally, we come to those varieties that have distinct vertical stripes on the skin. Until recently, the only striped color combination known was red with gold stripes. Recent work now provides us with green and gold, purple and green, yellow and pink, and other variations in between. The visual appearance of the striped varieties is simply stunning. Note that the special striped coloration is a characteristic of the skin and does not bleed through into the flesh. The best way to capture the unique beauty of the striped varieties is by using them whole for stuffing or in large wedges with significant portions of skin attached.

REPRESENTATIVE MULTICOLOR TOMATOES: Ruby Gold, Regina's Yellow, Hillbilly, and Lucky Cross (bicolored beefsteak types).

REPRESENTATIVE STRIPED TOMATOES: Green Zebra, Tigerella, Berkeley Tie Dye, Tiger Tom, Vintage Wine, and Copia.

TOMATO FLAVORS

If all tomatoes tasted the same, we would still be able to enjoy the wide varieties of shapes, sizes, uses, and colors. But what makes it all the more enjoyable is the amazing array of flavor nuances. To my palate, the most distinctive tomato flavor characteristics are tartness compared with sweetness, flavor fullness compared with blandness, and complexity compared with simplicity. Each of these characteristics is on a continuum as well, which makes the flavor combinations and impacts endless.

TOMATO FLAVORS

	Mild	Moderate	Intense
 Tart	Bonny Best Green Grape Green Zebra Tiger Tom Yellow Brandywine	Black Krim Black Prince Green Zebra Early Girl	Abraham Lincoln Jaune Flamme Old Brooks
 Balanced	Aker's West Virginia Andrew Rahart's Jumbo Red Ferris Wheel Persimmon Roma	Better Boy Big Beef Black from Tula Great White Kellogg's Breakfast Lemon Boy	Aunt Ruby's German Green Brandywine Cherokee Chocolate Cherokee Green Cherokee Purple Dester Green Giant Lillian's Yellow Heirloom Lucky Cross Nepal
 Sweet	Hugh's German Johnson Gregori's Altai Pineapple Ruby Gold Yellow Pear	Eva Purple Ball Yellow White	Mortgage Lifter Ponderosa Sun Gold

The Myth of the Low-Acid Tomato

Some tomatoes are marketed or described as "low-acid" varieties. Interestingly, a recent study disproved this myth by showing that the acid level in just about all tomato varieties lies within a very narrow range. Tomatoes that taste tarter simply possess less sugar, and those that taste sweeter are higher in sugar, which masks the sensation of acid. Some tomato varieties attack the taste buds in their assertiveness, while others are incredibly subtle, approaching blandness. Very few tomatoes possess perfect balance between sweet and sour, fullness and subtlety, and they can be fairly complex in flavor character. (If it sounds like I am describing fine wine, the analogy is actually quite apt. We find that tomato tastings are just as challenging, enjoyable, and varied as wine or beer tastings.)

Texture and Juiciness

Aside from flavor, several factors contribute to the enjoyment level of a particular tomato variety: texture, juiciness, tenderness or thickness of the skin, and relative ratio of seeds and surrounding gel to flesh. I've sampled tomatoes with very good flavor but flesh that is watery or has large empty spaces in the seed cavities, which left me with a negative impression of the variety overall.

THE EYE AFFECTS THE TONGUE . . .

People often presume that a correlation exists between color and flavor. I've attended enough tomato-tasting events and spoken to enough seedling customers to understand that many gardeners believe that red tomatoes have an assertive, acidic flavor. This could be because several of the older red-colored commercial varieties, like Rutgers and Marglobe, have a lower sugar content and therefore strike the palate as relatively tart. These types of tomatoes were the varieties of many people's youth. Similarly, lots of people think that pink, white, yellow, or orange varieties have a sweet, mild flavor. These are often mislabeled as "low acid" at farmers' markets. I've recently heard black tomatoes described as smoky or salty. If we were to blindfold tomato tasters, would they be able to name the color of the tomato being tasted? I suspect that the answer would be no. Through my years of growing and tasting tomatoes of every color, shape, and size, I can identify as many exceptions as general rules when it comes to correlating color and flavor. Since taste is specific to the palate of the taster, all tomato enthusiasts will have their own opinion on the matter. To me, it's a great reason to grow or purchase, and taste, as many varieties, colors, and types as you can. The only way to test this hypothesis for yourself is to find or grow the tomatoes to do your own tests.

TOMATO SEED TYPES

Which type of tomato to grow — hybrid, heirloom, or non-heirloom open-pollinated — depends on the expectations of the gardener. If your interest is primarily in growing a high-yield food-supply garden and you have no interest in saving seeds, focusing on hybrids is a reasonable approach. If your goal is great visual and culinary variety, open-pollinated and/or heirloom tomato types will provide everything you need — including the daunting task of selecting from thousands of varieties!

Hybrid

A hybrid tomato is grown from seed collected from a fruit that developed from a process known as *crossing*. Most simply described, pollen from one parent is directly applied to the pistil of another parent. Prior to the cross, the anther cone (the pollen-producing part) is removed from the receiving parent so that the flower doesn't self-pollinate as it typically would. If the ovary below the pistil swells following pollen application, forming a tomato, all the seeds in that tomato will be identical and are what is known as an F_1 (first filial generation) hybrid between the two parents. Tomatoes that come from those

seeds will show the dominant characteristics of each parent. Big Boy, created and released by Burpee in 1949, was the first famous hybrid tomato, and its popularity revolutionized the tomato-breeding industry. Recent examples of hybrid tomatoes are Better Boy, Celebrity, and Sun Gold. Often, the goal of creating a hybrid is to involve genetic material that helps a variety overcome tomato diseases.

Hybrids have a reputation for being more vigorous, more disease resistant, and overall, just plain better, wiser choices for the gardener. Often, what gets sacrificed in creating hybrid varieties is the single favorite goal for many gardeners: flavor. My own experience is that not all hybrids fight disease well, and certainly not all heirlooms succumb immediately. There are good reasons for commercial operations to focus on hybrids (such as a more concentrated harvest that works well for machine harvesting, specific disease resistance for a particular area, or smooth, uniform fruit for a particular market) but far fewer reasons for home gardeners to do so. The variety of color, shape, size, and flavor is significantly limited with hybrid tomato choices.

Seeds saved from hybrids will germinate and produce tomato plants that produce tomatoes. What *type* of tomatoes result, however, is not at all predictable or reproducible. Depending on the magnitude of differences between the two parent varieties, it may not be worth giving up valuable gardening space and effort in growing out seed saved from hybrids, unless you are embarking on a dehybridization process to work toward creating a new variety, or replicating the hybrid itself as closely as possible. (This is discussed in greater depth in chapter 7.)

Open Pollinated

Also referred to as non-hybrid, open-pollinated tomatoes have stable genetic material, and seeds saved from open-pollinated varieties will (unless cross-pollinated by bees) replicate the parent variety. All heirloom varieties are open pollinated, but not all open-pollinated varieties are considered heirlooms. Open-pollinated varieties may exist unchanged for hundreds of years; Yellow Pear is an example. Or they could be the result of recently discovered mutations, such as Cherokee Chocolate, or recent breeding projects, such as those from the Dwarf Tomato Breeding Project (see page 150).

Heirloom

An heirloom is an open-pollinated variety that has history and value embedded within its story. The term is used very loosely, since a tomato with the word "heirloom" (or "heritage") attached to it commands more respect (not to mention a higher price tag in many cases). Often, you will find fairly recent open-pollinated varieties such as Green Zebra listed as heirlooms.

My personal guideline is based on the release date of Burpee's Big Boy (1949). This event represented a major change in tomato variety development by seed companies. Prior to Big Boy, companies sold open-pollinated varieties; going forward, the vast majority of new varieties were hybrids. Given this, to my thinking, heirlooms are open-pollinated varieties that originated before 1950, whether they came from families in the United States or overseas, or from commercial ventures. This also creates two categories of heirlooms: the true family heirlooms that may have interesting histories and occasionally less-developed backgrounds (such as Brandywine), and non-hybrid, pre-1950 varieties developed and sold by seed companies, such as Golden Queen.

Heirloom varieties are wonderful not only for their variety of colors, shapes, sizes, and flavors, or for the ability to easily save seed and pass them on to others, but for the fascinating histories that seem to be associated with many of them. In some cases — such as Mortgage Lifter — we know quite a lot about the history of a particular variety. In other cases, we know far less than we'd like to, and efforts to learn more are often met with frustrating dead ends. Still, a garden filled with heirlooms becomes a kind of living museum and a great tool for teaching and storytelling.

Those who think of the world of gardening as idyllic and peaceful would be surprised at the level of rancor, disagreement, and misinformation surrounding a simple tomato name: Brandywine. This is a tomato that grows on a very vigorous, tall plant with large potato-leaf foliage. The fruits are large, pink, and oblate and often show unattractive irregularities. Some seasons, Brandywine simply shines, while in others it brings disappointment in terms of yield. However, after growing over a thousand different tomatoes over the years, it is still Brandywine that I think of when I ponder the perfect tomato-eating experience. An authentic Brandywine has an unmatched succulent texture that melts in your mouth. The flavor enlivens the taste buds, with all the favorable components of the best tomatoes — tartness, sweetness, fullness, and complexity — in perfect balance.

Because of its popularity, numerous Brandywine "relatives" have been introduced over the years — some with family names attached (Glick, Pawer, and so on), or different colors (Yellow, Red, White, and Black). It's not clear whether or how these are actually related to the real Brandywine, and, sadly, we will likely never be able to completely untangle the complicated history of this tomato.

Looking back into old seed catalogs, the only use of the term "Brandywine" was for a tomato created and released by the Johnson and Stokes seed company in the late 1800s. I've reviewed color plates of the tomato and it appears to be red, medium-size, and with regular-leaf foliage. This indicates to me that what was known back then as Brandywine is actually what we know of today as Red Brandywine, itself a really fine tomato. When considering the location of the seed company (Pennsylvania), the likelihood of the name originating from the river or town of Brandywine is very good.

There are indeed older tomatoes that carry the same potato-leaf foliage and large pink fruit as today's Brandywine. In the late 1800s, the Burpee Seed company listed a variety called Turner's Hybrid, and Henderson Seed Company carried Mikado. A few other seed companies also listed large, pink, potato-leaf varieties with their own particular names. Could it be that Mikado or Turner's Hybrid was grown and passed around through the years, finally becoming renamed by someone in the Midwest as Brandywine? Anything is possible.

My source for the wonderful Brandywine that I began growing in 1988 is seed saver Roger Wentling from Pennsylvania. Roger obtained his version directly from Ben Quisenberry, a legendary seedsman whose story is well documented by the Seed Savers Exchange in *Seed Savers Exchange: The First Ten Years, 1975–1985*. Ben claims that he obtained the tomato from Dorris Sudduth, and that the tomato was in her family for many years.

YELLOW BRANDYWINE

BRANDYWINE

Yellow Brandywine

A listing in a late 1800s Henderson catalog shows the release of a new variety named Shah, a large, deep yellow variety with potato-leaf foliage, as a "sport" or mutation or selection from Mikado. Some tomato enthusiasts simply assume that Shah is a white tomato. My conclusion, rather, is that what was known back then as Shah is very much like, if not a direct relation to, what we grow today as Yellow Brandywine.

Black Brandywine

This is a recent release by the Tomato Growers Supply Company. I grew it the first year that it was available, and it proved to be unstable, providing both regular-leaf and potato-leaf plants. Since its release, gardeners who purchased seed carried out selection work; I now have a potato-leaf selection that produces a very good medium-size purple tomato. The relationship between Black Brandywine and Brandywine is unclear.

GROWTH HABIT

The growth habit of tomatoes is quite important to the planning process, especially if your ability to grow tall, sprawling varieties is confined for various reasons, whether because of physical limitations or space or garden location considerations. There are, of course, nuances and subtypes, but for the purposes of this book, describing the three main types — indeterminate, determinate, and dwarf — covers the vast majority of tomato types typically available.

Indeterminate

This class of tomatoes is by far the most common, and its members grow upward and outward continually until killed by frost or disease. Indeterminate tomatoes have a central main stem from which side shoots, or suckers, grow at a 45-degree angle outward from the attachment point of the leaf stems. In turn, each sucker or side shoot acts as an additional main stem and produces its own side shoots. The central stem of an indeterminate tomato, if vertically staked and tied, will easily exceed 10 feet by the end of the growing season. Flowering clusters appear at varying intervals along the main stem and side shoots, ensuring continual fruit potential until the plant dies; this allows for continual and extended harvest throughout the growing season. Another important characteristic of indeterminate varieties is the relatively high ratio of foliage to fruit, and all of that added photosynthesis means a significantly higher flavor potential when compared with determinate varieties. There are many options for supporting and pruning indeterminate tomato varieties (see chapter 4). Well-known examples of indeterminate tomatoes are Cherokee Purple, Sun Gold, and Better Boy.

Determinate

Determinate varieties are far less common than indeterminate varieties; the gene that produces determinate growth habit didn't appear until the 1920s. Determinate varieties look identical to indeterminate varieties as young seedlings,

Indeterminate varieties, when supported, will grow upward and outward until killed by frost or disease, leading to potentially huge and unruly plants.

Determinate tomatoes, such as Roma, top out at 3 to 4 feet tall and are well controlled by cone-type tomato cages.

with the same stem width and foliage shapes and textures. There is a genetic component, however, that signals an end to vertical growth, emergence of flower clusters at the end of flowering branches, and massive fruit set over a very concentrated time span. This leads to a very narrow window for fruit ripening, which makes determinate varieties very attractive for commercial ventures that benefit from picking the fruit in just a few rounds of harvest. Because of the way the flowers appear, any pruning of this type will significantly reduce yield. In addition, the very high ratio of fruit to foliage means less photosynthesis; as a result, the vast majority of determinate tomato varieties have less flavor intensity and potential than indeterminate varieties (though there are always exceptions). Because of their compact growth, determinate tomato varieties are perfect for container gardening and caging. Well-known examples are Roma, Taxi, and Southern Night, as well as the Mountain series of commercial hybrids.

Dwarf

This rather unique class of tomatoes is largely unknown to most gardeners, although its profile has been raised recently from the efforts of our Dwarf Tomato Breeding Project. Dwarf tomatoes predate determinate varieties by decades, but very little development work seems to have taken place with them. They are defined by very specific and unusual growth habits. As young seedlings, the stems are very thick and far shorter when compared with those of indeterminate and determinate tomato seedlings. They retain this characteristic throughout their lifetime, behaving like slow-growing indeterminate tomatoes. The foliage is very dark, nearly bluish green, and quite crinkly looking; the term for this is *rugose*. The ratio of foliage to fruit is quite high, meaning that dwarf varieties possess a greater potential for excellent, full flavor than determinate varieties.

Prior to the advent of our Dwarf Tomato Breeding Project in 2006, the number of dwarf tomato options was limited to just a handful of old, historic varieties such as Dwarf Champion, along with a more recent creation, Lime Green Salad. As of 2013, the Dwarf Project has added another 17 varieties to the mix, and many more are yet to come.

Dwarf tomatoes are able to thrive in containers as small as 5 gallons. They top out at only 3 to 5 feet by the end of a long growing season, which removes the need to stake and deal with rampant vines. They can also be used in close-planting schemes in the garden, raising the yield potential per square foot. Among the new and exciting dwarf tomato releases are Dwarf Mr. Snow, Dwarf Sweet Sue, Summertime Green, and Rosella Purple. The Dwarf Tomato Breeding Project is described in much greater detail in chapter 7.

Dwarf tomato varieties behave like indeterminate varieties in fruiting but grow vertically at about half the rate. They are perfectly happy in 5-gallon containers and are very easy to control with short stakes or cages.

MANY FORMS OF FOLIAGE

Knowing which type of leaves a particular variety should exhibit can help in spotting crossed or mixed-up varieties in the garden. Most tomato varieties have regular leaves with serrated edges. A few well-known heirlooms, such as Brandywine, show the recessive potato-leaf foliage (which is smooth-edged). Dwarf tomatoes have very dark blue-green, crinkly (rugose) leaves, which appear in both regular- and potato-leaf variations. Heart- and paste-shaped tomatoes tend to have droopy leaves. Wild types, such as Mexico Midget, have very small, finely divided, delicate foliage.

LEAF SHAPE

Tomato leaf shape is not so much a parameter for planning as basic information to enhance understanding of the genetics of tomatoes, and it provides one additional characteristic used to distinguish varieties. Most people who grow tomatoes easily recognize "regular-leaf" foliage, with its characteristic toothed edges. This is a dominant trait that is represented by the vast majority of tomatoes, including Better Boy, Sun Gold, Celebrity, Roma, and countless others.

Potato-Leaf Foliage

Potato-leaf foliage is far rarer. Leaves of this type have smooth margins, with a strong resemblance to the foliage of potato plants (hence the name). This type of foliage is a recessive trait. An easy way to test the success of seed saving with regard to purity is to examine seedlings grown from potato-leaf plants that were themselves grown adjacent to regular-leaf varieties. Any regular-leaf plants that show up during germination will most likely signify an insect-induced cross. It is interesting to note that even though potato-leaf foliage is relatively rare, many famous and delicious heirloom varieties, such as Brandywine and Lillian's Yellow Heirloom, have this type of foliage.

Rugose Foliage

Rugose foliage is specific to dwarf tomato varieties. Dark bluish-green, rugose foliage is highly folded and crinkled, and particularly lush and attractive looking. Rugose foliage can have either the regular (serrated) or potato-leaf (smooth edge) shape. Such foliage makes the dwarf tomatoes particularly striking plants.

By far, most tomato varieties will have serrated or toothed leaf edges, which is the dominant foliage trait. They are known as regular-leaf varieties.

Tomato foliage with smooth edges exhibits a recessive trait known as potato-leaf foliage, since there is a resemblance to the leaf shape of potato plants.

MEET THE VARIETIES: CHEROKEE

Cherokee Purple came to me as an unexpected gift one day in 1990. It arrived in the form of a letter and packet of seeds from John D. Green (he calls himself "JD") from Sevierville, Tennessee. He could not have imagined what would happen after I grew the seeds he so generously sent. All we know of the actual history is that JD's neighbor shared the seeds with him, and that they descended from a purple tomato given to the neighbor's family by Cherokee Indians about a hundred years ago. As is often the case, I wish I had spent more time asking Mr. Green more about the background of the tomato — the neighbor's name, for example.

The seeds arrived in time for me to include a plant of the yet-to-be-named "purple" variety in my garden that year. I loved the story that JD told but was skeptical about the color designation; all known tomatoes up to that date with "purple" as a descriptor ended up being pink. Also, remarkably, my garden that year contained Black Krim, the very first "black" variety.

The unnamed purple and Black Krim were definitely among the most interesting varieties I grew that year. To my amazement, ripening brought forth a color that I'd never seen before: a deep, dusky rose color that shaded to nearly true purple at the shoulders. Both the purple variety from JD and the Black Krim showed the same color. I was very excited to be observing something quite new in tomato color; of course, there has been a veritable explosion of "black" varieties since the early 1990s.

Perhaps those two new varieties looked the same, but when sliced and tasted, they were as different as could be. Where Black Krim had a distinct absence of sweetness — in fact it tasted quite flat and lifeless to my wife, Sue, and me — Cherokee Purple exploded in our mouths in a symphony of flavors and nuances. We loved it (and do to this day). Clearly needing a name, I considered the brief history proposed by JD and named it Cherokee Purple, also listing it in the upcoming SSE yearbook and sending a sample to Jeff McCormack of the Southern Exposure Seed Exchange. Jeff called the following autumn and related to me his admiration for the flavor, but also his concern about whether the public would accept such an unusual color "resembling a bruise."

Years have passed and it is clear that the unique color of the "black" tomato varieties, with the genetic disposition to retain some of the green of chlorophyll upon ripening (thus leading to the deep crimson flesh color and dusky, purple-shaded outward appearance when viewed through the clear skin), has been enthusiastically accepted by gardeners and chefs everywhere.

I've spoken to Mr. Green through the years, and he is excited about how popular Cherokee Purple has become, and the last time he and I spoke, he still grew it annually. Unfortunately, he couldn't shed any additional light on the history of the tomato and wasn't able to provide the name of the neighbor with whom the Cherokee Indians of the area supposedly shared the seed. This is not an unusual aspect of heirloom tomato stories; we frequently have tantalizing tidbits, but attempts to dig further lead to dead ends.

Cherokee Chocolate In 1995, I was growing a number of Cherokee Purple plants from saved seed. I noticed something unusual about the way the fruit on one particular plant was coloring up as it approached ripeness. Though possessing the same darker-than-typical hue of the "black" varieties, the tomatoes showed something completely different: rather than the dusky rosy purple hue, the ripe outward appearance was more like a brick red or mahogany, and it was just beautiful to behold. The plant habit was exactly the same as that of Cherokee Purple, as was the size and shape of the fruit. In fact, when the first ripe fruit was cut, in its internal structure and color — indeed, even its flavor — it was identical to Cherokee Purple. Clearly, the differing characteristic was the skin color. In this new find, a yellow skin over the dark interior was responsible for the unique chocolatey color.

There were two possible causes for what I observed. If I was fortunate to have planted a seed with genetic material possessing a mutation for skin color, from clear to yellow, seed saved from the newly colored tomato would predictably produce the chocolate "form" of Cherokee Purple. If, instead, the seed I grew was from a fruit resulting from a bee-produced cross pollination, there would be an array of possibilities in the fruit grown from seed saved from the brown fruit. The following season I was delighted to find that all plants produced from the seed I saved from the uniquely colored fruit produced the chocolate-colored variant. I was very lucky to have grown that seed, and a new variety was born, which I called Cherokee Chocolate.

Cherokee Green The story continues, but in an even more unusual way. After I listed Cherokee Chocolate in the SSE yearbook, a seed request came in from a tomato enthusiast named Darrell Merrell, well known in Oklahoma as "The Tomato Man." Darrell ended up growing my seed, obtained the expected brown-colored fruit that now defined Cherokee Chocolate, stated how much he enjoyed trying out this new variety, and sent me some seeds that he saved. I grew some of those seeds in 1997.

Upon looking over my tomato patch that year, I saw something unusual going on with one of the Cherokee Chocolate plants grown from Darrell's seed. While all of the other tomatoes in the garden were turning their expected ripe colors, that one plant set lots of tomatoes of the right size and shape, but they were not turning chocolate — or purple, or anything else for that matter. When I looked closely, they did seem to be getting a rather unusual and unexpected yellowish cast. I squeezed a few of the fruits and was surprised to find that they were dead ripe. Picking one of the fruits and cutting it open completed the surprise and brought my delight. It was as green as could be, but one taste told me that it had the same scrumptious flavor shared by Cherokee Purple and Cherokee Chocolate. So, once again, I was faced with a mystery and a problem to solve.

Like Cherokee Chocolate, Cherokee Green has yellow skin, but its flesh color shifted from red to green. If the shift came from a genetic mutation, it would last through successive generations. If it came from an inadvertant cross, other colors would start showing up immediately. The results here are not so clear-cut. Over time, most of the people I've shared Cherokee Green seed with did get the green flesh and yellow skin; however, some occasionally found other colors in the mix. It could just be that we are observing an instance where selection in a cross for a recessive trait, such as green flesh, is easier to stabilize in less generations of trial. In the future, I hope to get a better handle on Cherokee Green in terms of confirming whether it is indeed from a cross or a mutation.

Whether it was initially a cross or a mutation, through years of growing, seed saving, and selection, it is now a stable, excellent open-pollinated variety. In a way, both Cherokee Chocolate and Cherokee Green are examples of how new varieties have been identified, shared, and stabilized throughout tomato history.

Planning AND PLANTING

SUCCESSFUL TOMATO GROWING is not difficult and there are many ways to do it successfully. It is, however, more involved than planting a six-pack of marigold seedlings around the front mailbox (though just how much more involved depends upon your goals). This chapter will lead you through many options, from very simple to more complex, and will help you choose your own particular journey.

PLANNING FOR SUCCESS

Start planning your garden in the winter, as the seed catalogs begin to arrive in the mail. Good planning up front is a significant step toward ensuring a successful garden. The level of detail in the garden plan is up to each gardener. After the initial round of planning, you may want to consider what comes next. For example, those with longer growing seasons may consider a midsummer planting in order to extend the harvest. If you have a relatively short growing season for tomatoes, careful planning to ensure that the plants are hitting their stride during the optimal temperatures is most important.

One of your first planning decisions is whether to buy seedlings or start your own from seed. If you have the time, space, and desire to start your own seed, your tomato options are endless, in terms of both which tomatoes and the source of the seeds. Though starting with seedlings means your options are far more limited, the good news is that each year brings a wider selection of available varieties. You must consider whether the options for purchased seedlings will satisfy your needs, and whether you're willing to take on the task of starting the tomatoes from seed.

There is, of course, great satisfaction in producing your crops by starting from seed. The period between starting the seed and planting out the seedling is filled with drama, impatience, and anticipation, as well as unanticipated surprises. Though all of that is eliminated if you choose to purchase seedlings, for the vast majority of gardeners the true joy comes during harvest and consumption. My intent is not to favor one path or the other but to provide adequate guidance to help you succeed no matter which path you choose.

STARTING YOUR OWN PLANTS FROM SEED

One of the most difficult parts of planning to start from seed is simply selecting from among the incredible array of available varieties. There are many good sources for tomato seeds, the main ones being seed catalogs (from both the major players and the smaller specialty companies), various Internet sources (eBay, Craigslist,

Thick planting of tomato seeds is a great way to produce lots of plants in a very small space. Adequate light is critical to proper growth, and shop lights hung so that the bulbs are within a few inches of the top of the growth will result in vigorous, stocky seedlings.

online-only companies, tomato and garden discussion group message forums), seed sharing organizations (the most significant of which is the Seed Savers Exchange in the U.S.), and racks at all sorts of retail outlets, as well as trades with other gardening friends.

DO YOUR RESEARCH

It's important to have confidence in your seed source. The recent popularity of heirloom vegetables and seed saving is both a blessing and a curse. In order to create specific buzz and elevate sales, some heirloom tomato stories end up being fabricated or embellished to the point of utter inaccuracy. Often, labels such as "hand selected" or "organic" are simply an excuse for the supplier to ask ridiculous prices for a tiny quantity of seeds. Do some research before purchasing from the many smaller, newer tomato seed sources to ensure that you are dealing with reputable companies providing reliable information. On page 234, I've included a list of the tomato seed suppliers I recommend, based on many years of research and personal experience.

YOU CAN DO IT!

Some gardeners lack the confidence to start their own tomato seedlings. My goal as a supplier of seedlings has always been to teach our customers so well that we will eventually put ourselves out of business. The main thing I've learned from years of gardening with our own seed-started seedlings is how easy it really is. But I've also learned that some steps are absolutely critical to success. Each season, a handful of our customers bring us the same story. They ordered a lot of seeds and got them going, and then disaster struck at some point, so they ended up buying seedlings from us. I hope that after reading through this chapter, you will have the confidence and information to provide for yourself all of the healthy seedlings you need for a great tomato garden.

Success with Seed Starting

The main factors for success in tomato seed starting are:

- Plans and a record-keeping system that are appropriate for the scale of your effort, including an understanding of key dates for your zone/climate
- High-quality seeds
- High-quality planting mix
- Flats or containers of a type appropriate for your needs
- Appropriate conditions for starting the seeds (location, temperature, sun exposure, moisture)
- Time and attention at different points of the process, as required

THE ONE-MONTH RULE

Perhaps the most important planning consideration is deciding when to start tomato seeds. Beginning on page 224, you will find data collected from tomato enthusiasts across North America that describe timing of seed starting, transplanting, and final planting out into the garden. My own guideline (which I call the "one-month rule") is to work backward from the plant-out date, which is typically when you can be sure there will be no more frost. If I plan to plant seedlings into the garden on May 1, the young seedlings are transplanted from small cells to larger pots on April 1, which means that my seeds are started on March 1. This timing works well for me here in USDA Zone 8.

QUALITY SEEDS

Start with seeds that have been stored properly or purchased (or bartered) from a reputable vendor, seed saver, or fellow gardener. If the saved seeds are older than 10 years, you may want to plant more than you normally would.

PROPER PLANTING MIX

In the world of cooking, they say the secret to a great recipe is the quality of the ingredients. The same applies for starting seeds: success relies heavily on the quality of the material into which the seeds are planted. In my experience, the very best material to use is a sterile soilless mix that moistens easily and stays fluffy. The main components of a good seed-starting mix are peat moss, perlite, bark ash, and dolomitic limestone, typically with a trace of starter fertilizer and a wetting agent. Inexpensive bags of heavy garden soil — or worse, your own garden soil (especially if it is heavy with high clay content, hence poorly draining and slow to dry) — are an invitation to disaster. Soil from your garden could also be infected with fungal, bacterial, or viral organisms that can lead to problems all along the way, from early damping-off of young seedlings (caused by a soilborne fungus) to infection of plants with diseases such as fusarium wilt. Also, your newly sprouted seedlings will be much happier if the tender, developing root system is growing in a medium that is porous, so that oxygen uptake can occur. Seedlings can quickly drown in heavy, waterlogged soil. This is not the place to skimp on quality, since it is the very beginning of your season's garden.

FLATS OR CONTAINERS

The options are endless for containers to hold your precious planting medium and emerging seedlings. Any container that can drain from the bottom will work, including cleaned-out yogurt containers, various-size plant pots, flats with cells of various sizes, and egg cartons. The key is cleanliness, since dirty containers can easily introduce disease. If you are using pots from previous years, soak them briefly in a solution of 1 part household bleach to 9 parts water, or a dilute trisodium phosphate (TSP) solution.

THE RIGHT CONDITIONS

LIGHT. Direct light (whether artificial or from the sun) is not essential for germinating tomato seeds. Once the seedlings emerge, light becomes critical. If it is the best that you can manage, a sunny window will provide adequate light for young tomato seedlings. Be sure to turn them once a day to counteract the process of stretching toward the sun (phototropism). Note that even disciplined turning won't prevent the seedlings from becoming leggy, so adjust your seed-starting timing so that a move toward hardening off and outdoor sun can soon follow.

However, if possible, the best way to give seedlings enough light is a simple, inexpensive shop light fitted with fluorescent tubes, which can be suspended just above the tops of the seedlings and raised or lowered as needed. Plants need rest time, so be sure to turn off the lights at night. I run my grow lights from approximately 8 a.m. until 10 p.m., but there is no magic to these exact times; the lights are on when I am awake and are turned off when I head to bed.

TEMPERATURE. Soil temperature between 70 and 80°F will ensure good germination.

A freely draining planting mix, such as a commercial soilless mix amended with compost, is perfect for growing tomatoes in containers. The fact that such plants will never become waterlogged is offset by a need to water vigilantly when the plants get big and weather becomes hot.

One useful tool here is a heat mat, which keeps seed flats at a constantly warm temperature. Using a heat mat really speeds things up; my tomato seeds normally germinate within 3 to 4 days when I use one. The mats also help mediate temperature swings that may occur in your seed-starting location.

MOISTURE. Even soil moisture is also important. A mix that's too wet will promote disease and is likely to result in the seedlings damping-off soon after they germinate. If the mix is too dry, the tiny seedlings won't receive enough water to live and they'll quickly shrivel and die. Placing a tent, dome, or other loose material over the flats of emerging seeds can help with moisture retention, but I find it important to remove it occasionally to allow for adequate air flow during germination, and permanently once the majority of seeds germinate.

SITE. As far as where to carry out the seed-starting process, consider the light, temperature, and moisture needs and see what works for you. I start my seeds on a table in my office, in front of a south-facing window. Other gardeners put the flats on top of a refrigerator or near a clothes dryer to provide warmth. Once the seedlings germinate, strong light, from either the sun in front of a window or grow lights, is essential; more detail is provided on page 71.

TIME AND ATTENTION

It is important to consistently monitor the conditions and progress of your seed-starting setup. If you have a good setup with appropriate materials, you'll be able to grow seedlings well with check-ins every few days before seedlings emerge. Following that, daily examination will help ward off trouble if it starts.

HARDENING OFF

Young tomato seedlings are very sensitive to cold temperature. The act of conditioning seedlings over time from tender and vulnerable to robust and strong is called *hardening off.* Tomato seedlings should never be moved from the warm indoors under grow lights directly to the final garden or container location without taking them through the gradual hardening-off process.

I harden off seedlings at two different points: before transplanting, when they are still in their small growing cells, and following transplanting into individual containers. The reasons for each of these steps are different. The initial hardening off, when the seedlings are still very small and tender, is to get them into direct sunlight as soon as possible to promote stocky growth and minimize stretching toward the sun (which results in leggy seedlings). The second hardening off is to allow seedlings to recover from the root damage associated with transplanting and is more about keeping them out of direct sunlight for a few

TOMATO SEED LONGEVITY

One important fact about tomato seeds that is not widely known is their longevity. In my experience, tomato seeds, whether purchased (hence a new crop) or freshly saved, when kept dry and away from temperature and humidity extremes, will germinate well for a minimum of 10 years. I find that even at 14 years old, enough seeds will germinate to produce decent plants. Something seems to happen between 14 and 16 years that renders the seeds mostly or completely dead or dormant. I mention this here because tomato seeds often comes in packets with at least 20, and often up to 200. Knowing that the seeds will be good for some years is comforting, and if you are not crazy about the variety after trying it out, the remaining seeds can be shared or traded with other gardeners.

days until the water-drawing ability of the roots recovers. Allowing newly transplanted seedlings to recover outdoors in the shade, or in a cool garage, for at least a couple of days will lessen the shock of the transplanting process. See page 72 for more details.

A stubbornly stuck seed coat, as shown here, is known as "helmet head." This is most often seen when germinating seedlings become too dry or the seed was old.

ONGOING CARE AND TROUBLESHOOTING

WATER WELL. Be sure that the planting mix has adequate moisture; newly emerged seedlings have shallow roots and can dry out quickly. Let the surface of the planting mix dry out between waterings, but watch for plants that start to wilt — a sure sign that watering is needed at once.

There are two ways to water flats or containers of young seedlings. The first is to place the containers into a larger container that holds water, and allow the water to be absorbed through the bottom drainage holes (this is often called *bottom watering*). Alternatively, you can water carefully from above, using a watering can with a narrow neck, ensuring that the foliage doesn't get wet.

WATCH FOR "HELMET HEAD." If seedlings emerge with the seed coat still attached, spray them with a bit of water each day to soften the seed coat. If you can carefully remove it with your fingernails without damaging the growing tip, do so. In some cases, the helmet head will not release and will strangle the seedling; this happens most often with very old tomato seed.

To prevent damping-off of seedlings (a fungal condition, as shown above), thoroughly clean all seed-starting equipment in a 10% bleach solution and provide adequate air circulation after seeds have germinated.

LOOK OUT FOR DAMPING-OFF. The main issue faced by seed starters is damping-off of very young tender seedlings. Damping-off happens when various types of fungi attack the seedling stem near the soil line or just a bit higher up. The stem pinches off the flow of water and nutrients to the top of the plant. The spores that become fungi that cause damping-off may be present in the containers (if reused), growing mix, water, or air. Using new containers and new, sterile seed-starting medium will go a long way in

minimizing damping-off. Since I've standardized the method outlined in detail in this chapter, I see essentially no damping-off issues, even when starting more than 5,000 seedlings each year.

NO FERTILIZER NEEDED. Please note that I haven't mentioned feeding or fertilizing newly emerged seedlings. Mother Nature packs everything needed into a seed for successful germination under appropriate conditions. Beyond that, proper growing mix has sufficient starter food to sustain vigorous early growth. Water, air circulation, and good light exposure, preferably real sunlight, as early as possible, in appropriate temperatures seem to take care of the needs of the young seedlings just fine.

START THE MOVE OUTDOORS. Once the seedlings have developed true leaves between and above the two cotyledon leaves, start looking for days when the weather is well above freezing. Ease the developing plants into the sun outdoors, paying attention to watering needs. Of course, this is strongly dependent on your planting zone. It is not uncommon for enthusiastic (read "impatient") gardeners to get things going too soon and find themselves with rapidly growing tomato seedlings indoors and inhospitable temperatures outdoors!

Weather will affect the apparent quality of seedlings, though damage is typically temporary. Very young, tender seedlings do not do well in high winds, especially if accompanied by heavy rain, before they are fully hardened off. And, of course, the enemy of tomato seedlings, and any young plant, for that matter, is frost. Any indication that seedlings (even hardened ones) could experience temperatures at or below 32°F calls for a hasty move to safety. Loose covering with a very light air- and water-permeable material, such as floating row cover or Reemay, will typically provide a few degrees of protection.

THE DENSE PLANTING TECHNIQUE

Of course, most gardeners have small gardens and have no need for hundreds of seedlings of dozens of varieties. My own plantings were quite small and limited until we decided to sell heirloom tomato seedlings at the Raleigh State Farmers Market some years ago. Without a greenhouse, we had to come up with a seed-starting method that allowed production of a large quantity of tomato seedlings of many varieties in a very small space.

Over the years, I've devised a dense planting procedure that works very well. I have no doubt that variations on the dense planting method have been used for decades by commercial ventures as well as creative home gardeners. Though best for those who wish to start a lot of plants, the principles can be applied on a much smaller scale. The informative video of this method that I made and posted online has received many views and positive feedback.

Up to 50 seedlings can be successfully grown in cells no larger than a few square inches. Dense plantings like this one need to be watered much more frequently, however.

MEET THE VARIETIES
MORTGAGE LIFTER

For many heirloom tomato enthusiasts, it is the delightful story of the creation of Mortgage Lifter that captures the imagination. Considering how many tomatoes are now available, it may be surprising to realize that only a small percentage of them have well-documented historical information. We know quite a bit about Mortgage Lifter because its creator was around to discuss exactly what he did to make such a wonderful tomato. It is the story of Marshall Cletis Byles of Logan, West Virginia, and it takes place in the early 1930s.

With absolutely no experience breeding or growing tomatoes, Mr. Byles had a goal of developing the largest, most delicious tomato possible. His unorthodox breeding technique involved using German Johnson, Beefsteak, and unknown Italian and English tomato varieties as his starting point. He planted the tomatoes in a circle, with German Johnson in the middle, surrounded by several of each of the other three varieties. With a baby syringe, he applied pollen from the edge varieties onto flowers of the central German Johnson. Saving seeds from tomatoes that formed from the cross-pollinated German Johnson blossoms, he repeated the process above for several years, and once he was happy with his results, he sold plants for a dollar each, thereby paying off his $6,000 mortgage in six years; people came from hundreds of miles to purchase the plants as news spread of their excellence.

Mr. Byles shared the seeds of Mortgage Lifter with the Southern Exposure Seed Exchange, which was the first seed company to make them widely available, labeled with the full name Radiator Charlie's Mortgage Lifter ("Radiator Charlie's" is dropped in most seed catalogs today). Though this is the story we know of for Radiator Charlie's Mortgage Lifter, it is possible that such a name was also used for similarly large, pink, sweet varieties; the SSE lists some that appear to predate the creativity of Mr. Byles.

A number of "strains" or regional versions of Mortgage Lifter arose through the years. Two that I regularly grow are from James Halladay of Fairview, Pennsylvania, and from Charlotte Mullens of Summersville, West Virginia, who tells me that hers is the favored variety of her parents. Mr. Halladay notes that his "strain" was developed in the 1920s or 1930s and was obtained from the area of Ashland, Kentucky/Huntington, West Virginia, where it has been grown for several generations. This seems to indicate that other gardeners used the name "Mortgage Lifter" for their own slicer tomatoes.

From the different strains I've grown, the results have been essentially identical. The rather weak, spindly-looking seedlings develop into huge, vigorous plants that are highly branched. The tomatoes that form from the often large, marigold-shaped flowers are oblate, pink, and very large, growing easily to 2 pounds or more. Best of all, they share a sweet, intense, delightful flavor.

MULLENS MORTGAGE LIFTER

WHAT LEAVES TELL US

There is subtle variation in the appearance of young seedlings of certain different tomato varieties. Paste or heart-shaped tomatoes typically have significantly narrower foliage and weaker central stems that have an increased tendency to topple over. I've noticed that they are the least happy varieties in dealing with cold, windy, wet weather when very young. Sun Gold, the wonderful orange hybrid cherry tomato, shows its dislike for early cool, wet weather by looking just plain unhappy and weak, often with brownish foliage spots, until the weather warms up and growth becomes vigorous. A very few older beefsteak types, most notably Kellogg's Breakfast, and certain heart-shaped varieties such as Yellow Oxheart quickly develop browning and drying on the cotyledons and new growth that looks alarming but vanishes once the seedlings are transplanted and growing in full sun. The issue, which some have named tomato seedling "crud," appears to be genetic, since the particular varieties that develop crud are afflicted from every seed source from which I've obtained such varieties. It could be that certain types are less able to deal with low-light situations, and this could be related to the nutritional needs or processes of the variety. What I do notice is those varieties that are crud-prone end up being some of the most rampant and vigorously growing varieties in my garden.

The method involves planting up to 50 seeds in small cells. The most useful vehicle I've found for starting tomato seeds in this manner are the 50-cell, rigid, plastic plug flats with cells that are 1½ inches square. The entire flat has dimensions of 1 by 2 feet. Following germination, the seedlings can be easily separated and transplanted into individual pots with nearly 100 percent success. Though the production of up to 2,500 tomato seedlings in a very small space may sound far-fetched, you can trust me, as this is the seed-starting method that I've used for the past 15 years with consistent success.

SHOULD YOU TRANSPLANT?

If you used a seed-starting method that results in one plant per growing area (whether using flats of cells, peat pots, yogurt containers, pots, or some other container), transplanting is not absolutely necessary, as long as that single seedling is hardened off and ready for planting into the garden when the weather is right. Transplanting is a necessity for me because I use the dense planting technique.

Regardless of which method you used to start your seeds, though, most gardeners believe that there are benefits to transplanting seedlings. When you take a flat of leggy seedlings stretching toward the sun and plant them deeply into individual pots, roots will form all along the buried stems. Not only does this process correct legginess and create stockier seedlings, but the stronger root system will also enable the plants to better cope with the elements when they're planted out to the garden. Remember that early spring often involves unsettled weather, including some cold, wind-driven rain. Leggy tomato plants with weak, floppy stems could experience damage, including breakage, in such conditions, but the stocky seedlings that result from the initial transplant will handle the uneven weather far better.

When my seedlings approach 2 to 3 inches in height and have well-developed true leaves, I separate them and transplant each seedling into its own pot. I've successfully transplanted seedlings even prior to true leaf appearance. This is often the first point at which I can confirm leaf shape in my seedlings and spot any surprises. I consider transplanting to be an essential part of the life cycle of my plants, and it helps them to develop the robustness and strong roots they will need to thrive in the coming growing season. With practice, it can be done quite quickly and effectively, with few, if any, failures.

The transplanting step is another place where using high-quality, light, fluffy soilless mix pays off handsomely. The same soilless mix can be used for both seed starting and transplanting. You will find that the closely planted, intertwined seedlings come apart very easily, with minimal breakage of roots. But you will also discover that some roots will break off, at no real detriment to the health and survival of the seedlings.

AFTER TRANSPLANTING

Once all of the seedlings from a cell are transplanted, I record the number and confirm the leaf type (noting any leaf types that are unexpected, such as potato-leaf seedlings in a regular-leaf variety), which allows for calculation of a germination percentage, if desired.

I typically have extra seedlings that are smaller than the ones I just transplanted. Rather than toss them, I fill a single pot with dry mix and gently push them into the mix, gathering them into bunches of two to five seedlings, if needed, to fit them all in. By treating straggler seedlings this way, even the smallest seedlings, over time, will grow into fully productive, full-sized plants. My pots of extra seedlings become my backup plants, and will require an additional transplanting step once they're large enough.

Freshly transplanted seedlings need time to recover from the disturbance that they just experienced. Putting out newly transplanted seedlings — even if well watered — into full sun leads to significant loss, because the slightly damaged root systems can't pull sufficient water into the seedlings to compensate for the hot, drying sun. However, after just two or three days of allowing the seedlings to rest and recover on my garage floor, they are ready to be moved into the sun. I do watch them carefully for the first week and ensure that they are well watered. The seedlings will go through a period when they don't seem to grow very much at all, especially if the nighttime temperatures are cool (in the 40°F range). During this period, most of the progress is happening with the roots, in the soil and out of sight. A few nice warm, calm days and mild nights will trigger very rapid growth, eventually leading to lovely-looking, healthy, and vigorously growing seedlings.

STARTING FROM PURCHASED PLANTS

There are many reasons to start with seedlings, rather than starting your own seed. If you didn't plan ahead, don't have the desired seeds in hand, have germination issues with the seed, or don't have space or time, you'll want to purchase tomato seedlings for your garden. Fortunately, the popularity of heirloom tomatoes has done wonders for the availability of seedlings, whether from seed companies, small seedling suppliers, or local farmers' markets.

THE PROS

The advantages of using pre-started seedlings (purchased locally or through mail order or shared from gardening friends) generally focus on convenience, space constraints, and timing. Starting tomatoes from seed requires up-front planning, selecting and purchasing of seeds, and an appropriate space in which to grow the seeds. And packets of seeds often provide far more seeds for a variety than are required.

By eliminating the steps and materials involved in starting your own seed, planting your garden with purchased tomato seedlings could end up being more cost-effective. By obtaining just the number of plants you need,

An easy way to create additional plants is by rooting suckers or side shoots of healthy plants. This cutting of Cherokee Purple rooted quickly in a glass of water.

you avoid the conundrum of what to do with the numerous excess seedlings that often result from overly ambitious gardeners starting too many plants of too many varieties. You also avoid the annual angst of trying to nuture self-started seedlings along in a healthy state in the typically unsettled late winter and early spring weather.

THE CONS

If you do choose to plant your garden with purchased tomato seedlings, it is important to know the various risks. Tomato seedlings are quite tender and there is always the chance that mail-order plants will not arrive in good condition. The cost of seedlings, when compared to seeds, is significantly higher, especially when considering shipping costs. And, of course, you'll have a much narrower selection of tomato varieties if you buy seedlings than you would if you started them yourself.

If you purchase seedlings locally, you might still encounter plants that have been mishandled, underwatered, or infected with diseases. Seedlings that are greenhouse-grown may not have been hardened off properly. (On this note, larger is not necessarily better, as smaller, well-hardened seedlings will adapt more quickly to being planted out than leggy, less-hardened specimens.) It is important to watch for seedlings that have been only recently transplanted and have yet to develop an effective root system in the container.

Finally, there is always a chance that the tomatoes you end up with are not what you expected (this is a possibility with seeds, too). One of the inevitable downsides to the popularity of heirloom tomatoes is that not all the vendors that have cropped up recently take care to ensure the accuracy of their seed sources. This just reinforces the importance of seed-saving organizations and those avid heirloom variety guardians, for whom historical and genetic accuracy is paramount.

When shopping for seedlings, be sure to ask questions to determine the vendor's level of experience and familiarity with the varieties offered. Fortunately, good planting practices following seedling purchase will result in eventual success in all cases, save misidentified or diseased plants.

USE CAUTION WHEN PURCHASING SEEDLINGS

When buying seedlings, be sure to avoid seedlings that:

- Seem too large and leggy for their container size
- Have open flowers and fruit on plants that seem far too young
- Show blemishes or discoloration on the foliage or stems
- Are wilted, crowded, or generally unhappy looking

Gardening takes and handsomely repays the efforts put into it throughout the season, and

everyone should aim for the best start possible to their season. Starting with substandard, unhealthy plants will only bring disappointment and frustration.

Once you have decided where you will purchase your seedlings and which varieties to purchase, all that is necessary is to decide when you want your plants to arrive in order to meet the appropriate plant-out date for your climate.

That said, to have the healthiest, most productive tomato plants throughout the growing season, you will need a few basic requirements, no matter where you live: ample sunlight, an appropriate temperature range, a suitable growing medium (soil or potting mix), good drainage, sufficient water, a support system, and a source of nutrition. We'll talk about all that in the next chapter.

When buying a large seedling, be sure to find one like this — one that isn't crowded in its container and has healthy-looking foliage, a sturdy stem, and no flowers.

WHAT ABOUT GRAFTED TOMATO PLANTS?

A very recent phenomenon is the creation of grafted tomato plants. In such plants, the growing top of a favored (but not necessarily disease-resistant) variety, such as Cherokee Purple or Brandywine, is grafted onto a rootstock that is disease resistant. The idea is that when such a specimen is planted into diseased soil, the chances of survival are increased because the disease is unable to enter the plant through the root system.

It's unclear what the grafting phenomenon will mean for home tomato growing going forward. In the United States, where tomato grafting is still in its infancy, there are strong supporters, converts, and true believers alongside yet-to-be-convinced skeptics. Reported benefits of grafting are quite stunning, including significantly increased yields, greater plant health, and flowering and fruit set under less-than-ideal conditions, combined with equivalent or superior flavor. Some think of grafting as the most important advance in tomato culture in nearly a century. The jury is out, but it sure does sound promising.

Some of the grafting caveats include the considerably increased cost of grafted plants (compared to the negligible cost of growing your own ungrafted variety from seed), the care needed in growing the plant (the top of the plant that defines the variety, above the graft line, cannot be allowed to set roots), and the possibility of the specific targeted diseases you wish to avoid being replaced by other unanticipated diseases. Also, any disease spread by passing chewing insects, or by being splashed onto the foliage of the top growth, can potentially take hold and affect the plant as if it were a normal, ungrafted specimen.

TRACKING THE TOMATO-COLLECTOR'S GARDEN

My annual garden plans always start just after Christmas with some serious thought about what I want to grow that season. Because I have a huge tomato seed collection, my varieties (which now number over 3,000) are broken down into options for our seedling sales, those that need to be grown out to ensure viability (typically before they reach 10 years of age), new varieties that people have sent to me over the years but have yet to be grown, the must-grow favorites, and, perhaps, new varieties that catch my eye perusing the seed catalogs or the SSE yearbook.

Though it's certainly not for everyone, working on various tomato projects can be quite fun and really bring some mystery and excitement into the garden. Examples could be focusing on particular colors, comparative trials of newly released varieties, or growing out something unexpected that arose in previous gardens. I always find myself including bits and pieces of several small projects in each year's garden. This is admittedly a bit over the top, but tomato growing, especially with the plethora of heirlooms available, can become quite addictive, and gardens can expand rapidly from year to year. But garden planning should be simple and enjoyable, inspiring a foundation of enthusiasm and energy that will carry through all the stages of the growing season ahead.

Setting Up a System

The winter days used to plan the garden are useful for setting up a system for record keeping or tracking garden results. The level of detail in such a system should be appropriate for your efforts and goals. The options range from no system at all (buy plants, plant them, care for them, harvest) to an extensive record of data and information that can be used for future gardens. I start each season with a fresh spiral-bound notebook, which serves as my garden diary, planning guide, and place to write information as the season progresses. In my notebook, I draw diagrams showing the intended location of plants, which helps me to assess reasonable spacing requirements. At the same time, I start a fresh worksheet on my laptop where I record all of the information I collect during the season, such as dates for planting and harvest of the first ripe tomato, fruit color, flavor, size, and shape. (If you do find yourself keeping track of your efforts on the computer, be sure that your data is backed up.)

The Importance of Keeping Track

The value of the garden log is in the ability to understand the duration of the key steps along the way. In reviewing the data, you can improve each gardening season by ensuring that you are starting the right activities on the most reasonable dates based on your past experiences. This is how I developed the "one-month rule" described on page 57, which I use to estimate the schedule for key activities from seed planting to setting the seedlings into the garden.

Having a garden log is also valuable if you experience the unexpected. For example, what if a Cherokee Purple tomato plant that you grew from your saved seed produces red tomatoes instead of purple ones? By looking at garden maps, it may be possible to determine what went wrong, simply by identifying possible varieties the bees used to make the unexpected hybrid you are now growing.

Ten Tastiest

Polish

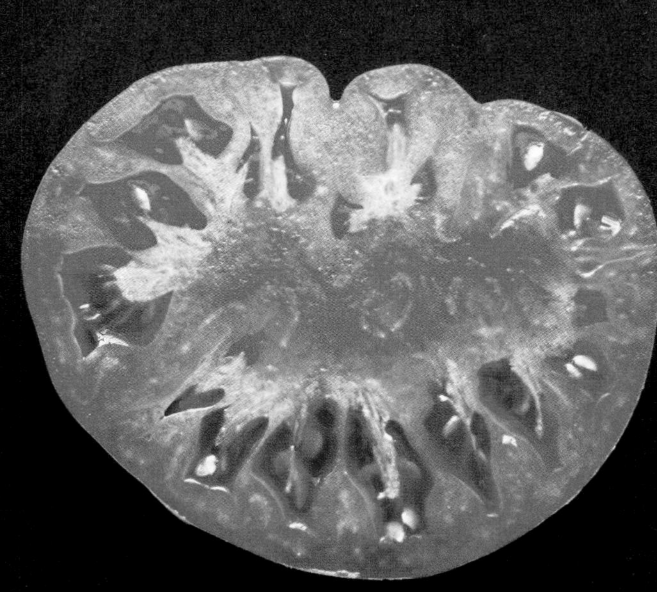

Before I discovered the joy of Brandywine, **POLISH** took a revered place in my garden for its productivity, beauty, and superb, richly flavored, wonderfully textured tomatoes. With the benefit of being a bit less temperamental from season to season than Brandywine, this large, pink beefsteak-type tomato is a regular in our annual gardens, and it is truly a tomato that has everything going for it.

SEED STARTING, STEP BY STEP

1. Fill the containers with dry planting mix to within ⅛ inch from the top. Gently water the container or individual cells with warm water until it drains slightly through the bottom.

2. Place individual tomato seeds onto the surface of the wet planting mix. It's up to you how many seeds you plant, from 1 to 50 for a cell that is 1½ inches square. This may be surprising, but it is absolutely fine to plant thickly and separate seedlings later on if you choose (see page 61 for a discussion of my dense planting technique).

3. Sift just enough dry mix over the seeds to cover them completely. Gently mist the top of the flat or container with water, using a fine sprayer. Loosely cover the container or flat with a fresh sheet of plastic wrap. Clear plastic domes can also be used.

4. Move the planted flats or containers to where they will spend time germinating. Light is not necessary, but warmth is recommended. Watch for signs of life after three or four days. Once seedlings start emerging, gently flip the plastic over and replace it each day to provide some air and to allow the condensation that builds up on the plastic to dry. Germination typically starts with the small loop of the emerging seedling showing above the surface, followed by the appearance of the first, smooth-edged "leaves." (Actually called *cotyledons*, or seed leaves, these are uniquely different from the true leaves that will follow.)

5. Once the majority of seedlings emerge, remove the plastic and ensure that the flats or containers get plenty of bright light. If you must, you can place the seedlings in front of a sunny window. Turn them daily to counteract the stretching-toward-the-sun process, a normal phenomenon called *phototropism*.

A better option is to set the seedlings in a cool spot, such as a garage, under grow lights (fluorescent tubes of any kind). Suspend the lights just over (within an inch) of the tops of the plants, raising them as the seedlings grow. Keep the lights on during the daytime and off at night. The combination of strong light and cool conditions will help the plants develop healthy, stocky stems.

PROJECT

TRANSPLANTING, STEP BY STEP

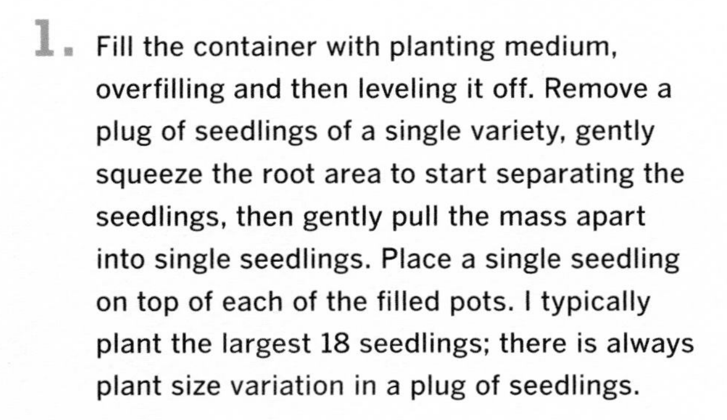

1. Fill the container with planting medium, overfilling and then leveling it off. Remove a plug of seedlings of a single variety, gently squeeze the root area to start separating the seedlings, then gently pull the mass apart into single seedlings. Place a single seedling on top of each of the filled pots. I typically plant the largest 18 seedlings; there is always plant size variation in a plug of seedlings.

2. Push each seedling gently but quickly and firmly into the pot until just the true leaves are sticking out of the soil, then pull the soil around the planting hole.

3. Using a watering can, gently but thoroughly water each of the pots until water starts to emerge from the bottom, focusing the stream at the soil surface so as to not wet the foliage.

4. Write the variety name on labels and then insert one into each pot. Place the planted flat on a garage floor or some other area of even, suitable temperature completely out of direct sun.

4

Growing, MAINTENANCE, AND Care

THERE ARE COUNTLESS WAYS TO grow tomatoes well. Rather than tell people what they must do, my preference is to share what has worked well for me, and then see what I can learn from others. My own gardening practices evolve over time; each season I learn new things. Many of the methods I employed living in Pennsylvania had to be abandoned for my Raleigh gardens because of different weather, soil, location, and tips I picked up through the years.

GROWING IN THE GROUND

Before raised beds, self-watering containers, and driveways full of pots, I suspect most people went into the yard, turned over a big rectangle of lawn, and created a garden. This is the way I grew tomatoes for many years, in rows or hills in a big patch of fenced earth that I enriched each year with additions of compost (or, even better, mushroom compost, which is well worth seeking out). I dug holes, planted seedlings, and pounded stakes into the ground to provide vertical support, and the result a few months later was the eagerly anticipated harvest. Since most of us plant tomatoes this way and have for years, I will focus on what I think of as the most important factors for success using this tried-and-true method.

Our family's first garden was in New Hampshire, in a college community plot that was created and maintained for students. It was wonderful — fenced, turned over each year, and given an annual, pre-season application of composted manure. Little did we realize how lucky we were to be able to easily stick a shovel into the rich, dark soil and not find rocks and clay deposits.

We were spoiled early on, that's for sure. All of our subsequent gardens had to be hand-dug. It's not hard for me to remember how much work it involved: cutting squares of sod, turning them over, chopping into smaller pieces, and finally tilling it all in. Our early gardens on any given patch of ground were very successful because of the plethora of untapped nutrients and absence of soilborne tomato diseases. We lived then in an area (west of Philadelphia) that received hard freezes each winter, so despite a limited opportunity to rotate the tomatoes into new areas of the garden each year, we saw no significant drop off in productivity or increase in disease issues.

Our experience in Raleigh has been quite different, with a yard that seems to ooze impermeable red clay and rocks. Aside from being a digging and drainage nightmare (issues we found ways to cope with effectively), the lack of a winter freeze, and therefore lack of a good mechanism for killing some stubborn pests, means declining success annually; it is this that caused me to move most of our tomato growing to containers. Our experience reinforces the need for each gardener to understand the local advantages and challenges of his or her own particular climate and soil conditions.

The leaves of autumn provide a perfect layer of mulch for the winter, slowly feeding the soil underneath and holding down weeds. Simply rake the leaves aside in spring to dig planting holes for the tomatoes.

The Planting Plan

Many enthusiastic gardeners tend to pack too many plants into a given plot. It's important to sketch out a planting plan that includes sufficient space for each plant to grow. The number of plants you can fit into a garden depends on the types of plants you plan to grow and how you'll grow them. Choosing dwarf varieties or training plants to grow vertically will allow for the densest spacing. Caging comes next, with the boundaries between plants set by the edge of the cage along with some space to allow for air circulation around the plants. Allowing the plants to sprawl takes up the most room, but of course it is the lowest-maintenance approach. In between are all sorts of variations — such as Florida weave, espalier, Japanese tomato ring, and training plants up extended strings or chains dropped from above (described on page 90) — each of which requires different spacing between plants to ensure the best success.

Check Your Soil Drainage

Walk into your garden, dig a hole at least 1 foot deep, and examine the soil structure. If you're lucky, the topsoil — the layer of dark, crumbly, worm-filled soil at the surface — will be at least a few inches in depth. Under this are the layers of subsoil of varying depths, which could contain rocks and clay. The structure of the subsoil affects the ability of the garden to drain and retain nutrients. Clay soil drains poorly but retains nutrients well; sandy soil, as you might imagine, does just the opposite.

You may find that the topsoil layer is quite shallow, and the clay-filled layer beneath is very dense and difficult to dig into, which poses a real drainage challenge. We didn't have to dig too deeply in our first North Carolina garden before we hit what seemed like brick material; it had very poor drainage. On the opposite end of the spectrum, you may find little to no topsoil, but simply a thick layer of sand with poor moisture and nutrient retention. No matter what you

IT'S TOMATO PLANTING TIME! OR IS IT?

The seedlings are ready to be planted, and you are already dreaming about what to do with your harvest to come. It is important, however, to take a deep breath, step back, and do some mental checks to determine the right time to plant. Though well-hardened tomato seedlings are reasonably robust, planting too early presents some real risks, and often no real benefit.

- **Check with your local Cooperative Extension agent to find the average last frost date for your area. After that, look at your long-range weather forecast. Tomato plants will die if they experience freezing temperatures, so 32°F is a very important number. If you have a relatively short growing season and want to take a chance, be sure to use physical protection such as water cloches or row-cover fabric if frost is predicted.**
- **Soil temperature and moisture content are nearly as important as air temperature. Seedlings planted in cold soil won't die, but they won't thrive either. And heavily waterlogged soil is likely to drown the young seedlings.**
- **In my experience, early-planted tomatoes that endure cold conditions sit and sulk, and those that are planted weeks later under more favorable conditions will catch up to the early specimens; soon, the seedlings from each planting, despite being weeks apart, will be indistinguishable.**
- **Garden centers always seem to push the season, and you will likely find seedlings far in advance of the recommended plant-out date. If you are seduced by particularly good-looking or well-priced seedlings, or you find must-have varieties, resist the urge to plant them out immediately; just keep them in good condition (give them adequate light and water, and perhaps take them through a gradual hardening off) until the time is right for planting.**

SPACING RECOMMENDATIONS

Vertically Grown

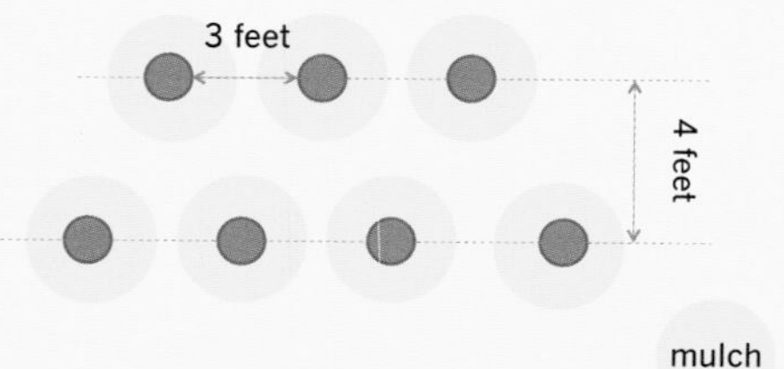

Vertically grown indeterminate tomatoes should have 3 feet between plants and 4 feet between rows.

Caged

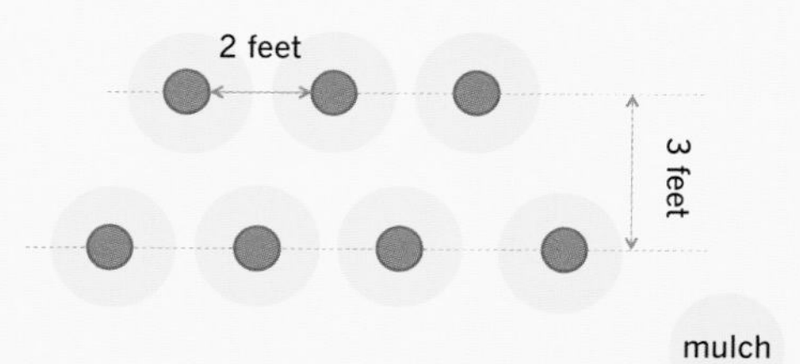

Cage-confined determinate or dwarf tomatoes should have 2 feet between plants and 3 feet between rows.

Sprawling

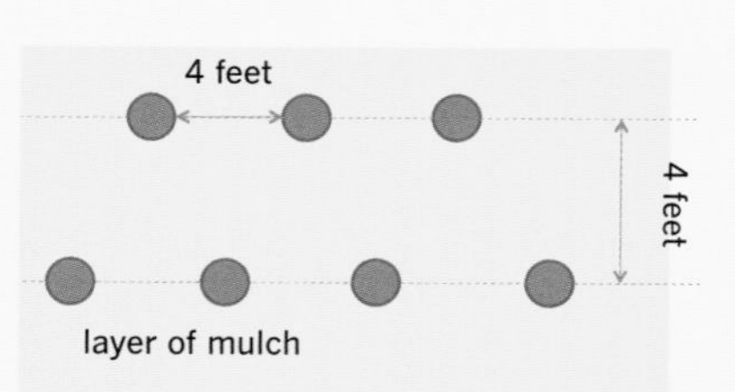

Sprawling indeterminate plants should have 4 feet between plants and 4 feet between rows. Note that the entire garden is well mulched to ensure that the soil does not come into contact with any plant parts.

observe, it's not difficult to improve the situation for your tomato plants.

After you examine the soil, fill the hole with water (or dig it before a significant rainstorm) and see how quickly it drains. If it empties gradually — within an hour or so — you're in luck: your soil structure allows for ideal moisture retention. Holes that remain quite full for a day or more indicate trouble and you should seriously consider planting in a different area or creating a raised bed (see page 83). Water that drains immediately indicates a very porous, sandy substructure; adding compost or leaf mulch will help considerably with the retention of water that your plants need.

Test Your Soil

Testing your soil for nutrient content (mainly nitrogen, potassium, and phosphorus) and pH level will be very informative and helpful, particularly if your garden plot has failed you in the past. Just a year or two of growing most kinds of vegetables in a garden plot is enough to deplete nutrients and necessitate a nutritional boost.

GETTING THE RIGHT pH

The relative acidity or alkalinity of soil is measured in a logarithmic scale called the pH scale, running from 0 (for extremely acid) to 14 (for extremely alkaline). Neutral pH, neither acidic nor basic, has a pH of 7. Tomato roots absorb needed nutrients from the soil when they are made soluble in a slightly acidic environment. So tomatoes do best in a slightly acid soil pH level of between 6.2 and 6.8. Soil that is too acidic will actually be deficient in some critical tomato nutrients, such as potassium and magnesium, because they will have leached out of the soil. If the acid is insufficient (with a pH above 6.5) the nutrients that are present will not be able to dissolve into a state that can be taken up by the plant root system.

The easiest way to adjust the pH of your soil so that it is fit for healthy tomatoes is by applying two simple substances. You can "sweeten"

WHAT'S IN YOUR SOIL?

Soil is composed of three main components: clay, sand, and silt. When these three components occur in roughly equal proportions, the soil is said to be "loamy." When one soil component dominates, conditions can be improved by regularly adding organic matter such as compost.

Clay

Soil dominated by clay is sticky when wet and rock hard when dry. Clay soils drain poorly and warm slowly but are often rich in nutrients.

Sand

Often a light color with visible particles of mineral and rock and gritty to the touch, sandy soil is easy to dig, drains well, and warms up quickly. Nutrients can wash through easily.

Silt

Silt is a granular substance made up of quartz and feldspar, with a particle size somewhere between sand and clay. It feels a bit like flour when it is dry but is quite slippery when wet.

Loam

Containing a balanced mix of sand, silt, and clay, loamy soil feels fine-textured and often a bit damp.

overly acid soil (make it more alkaline) by adding ground or pelletized limestone. You can acidify alkaline soil by adding sulfur. Rather than guessing on amounts, carry out a pH test using your local Cooperative Extension Service and follow the guidance of the results for the recommended adjustments. It is important to note the type of soil as well — sandy, loam, or clay, as described on page 79 — as this significantly affects the quantities of amendments needed to carry out the adjustment. For more about how to alter the pH of your soil, follow the links on page 235.

SUPPLYING NUTRIENTS

The most significant nutritional needs of tomatoes are for nitrogen (N), phosphorus (P), and potassium (K). Most gardeners are familiar with number designations on commercial fertilizers, such as 5-10-5, and this is what the numbers refer to — the relative amounts of N, P, and K. For example, for every 100 pounds of 5-10-5 fertilizer, there are 5 pounds of available nitrogen, 10 pounds of phosphorus, and 5 pounds of potassium. In general, tomato plants need relatively less N and relatively more P and K.

As an alternative to balanced, mixed (complete) fertilizers, you can provide each element individually. This is recommended if a soil test indicates the need to significantly adjust one of the components. Potassium (K) levels can be boosted with greensand or potash. Phosphorus (P) can be added by using bonemeal or rock phosphate. Animal manure, blood meal, and urea provide nitrogen (N). There are many other choices for organic approaches to fertilizing, such as using fish emulsion and seaweed preparations.

Compost

All soil can be improved by adding compost. In fact, any kind of organic matter that absorbs and holds water, or provides air spaces, will change things for the better. The options are numerous: composted animal manure (cow, horse, chicken, or rabbit), composted leaves, grass clippings, and homemade compost from kitchen scraps. The general guideline for soil is that it must drain well but hold water long enough for the plant to utilize. Whether the soil is too heavy or too sandy, adding organic matter in the form of compost will help. It's recommended that fresh manure not be used, as it could burn the roots

A greenhouse-grown seedling (like the one at left) will often be far leggier than a home-produced seedling (at right), so deep planting is recommended; roots will form along any part of the tomato stem beneath the soil line.

and harm the plants, though there are different schools of thought on this. When in doubt, compost the manure for a period of time prior to use.

If you don't have your own compost, the local hardware stores will have plenty of good bagged options, including various composts, manures, peat moss, and the so-called soilless mixes, which contain essentially the same materials as seed-starting media. Be aware that peat moss used in large quantity increases soil acidity. Often, an excessively acid soil will show itself when the tomatoes set fruit that start to enlarge and then show signs of blossom-end rot (BER).

Many tomato growers have their own ways of improving the tomato planting hole. Adding, from bottom to top, composted leaves, manure and straw, cottonseed meal, bonemeal, lime (as a preventive against BER if the soil is on the acid side), and sand, and topping with soil, would give the plant everything it needs to grow well.

One of the best mulches for tomato plants is chemical-free grass clippings. As they break down, they add nutrients to the soil; they also keep weeds from sprouting and prevent soil from splashing up onto the lower tomato leaves.

Early Planting Considerations

Tomato plants that are planted in cold soil and under the threat of continued cold weather, even frost, will not progress quickly at all and will be equaled by plants set in much later, once warmer weather arrives. If you must get your plants in early, because of either your particular availability and schedule or the desire to grow early tomatoes, you can use products or methods to protect the plant and provide some needed warmth. Season extenders include water cloches (such as the Wall-o-Water) and hot caps, as well as temporary physical barriers for larger garden areas, such as tunnels built of PVC hoops and covered with materials such as plastic or row cover. If you do plant out your tomatoes early and frost is forecast, you must provide protection. If you cover the plants with pots, be sure the foliage doesn't touch the sides. Row cover, loosely draped over the plants, offers a few degrees of protection, and several layers can be used. Plastic covering is not a good idea as it transfers the cold temperatures to any foliage that it comes in contact with

CUTWORM PROTECTION

Cutworms are annoying springtime pests that can encircle and cut through the tender stem at the soil line, severing the plant from the root system. Easily ward off cutworm damage by fashioning a loose-fitting collar (use paper or aluminum foil) around the plant base, with some of the collar lying below, and some above, the soil line.

and causes visible damage. To provide the most warmth possible for early transplants (especially in regions with a short growing season), try "trench planting" — near-horizontal planting at a shallow depth, with just a few inches of top growth exposed.

Mulching

Once a seedling is planted and the cutworm collar (if using) is in place, the next step should be mulching. Mulch options include untreated grass clippings, shredded hardwood leaves, newspaper, or any other layerable, water-permeable barrier that will keep the soil off the foliage. Many a tomato plant has become infected very early in the season by having microbe-filled soil splashed onto the lower foliage; mulching is the key way to avoid this. You can also use landscape fabric or plastic mulch. Research has shown that mulching with red plastic can significantly increase yield and speed up fruit maturation.

Deep Watering

The final step after planting is a deep watering at the base of the plant. It's a good idea to plant your seedlings in the late afternoon so that they have the nighttime to adjust a bit before being exposed to the full sun. If your tomato transplants are from individual pots with well-developed root systems, adjustment will be very easy. If you're separating tomato plants from a mass of seedlings, you'll have to watch them carefully for a week or so after planting, watering often to ensure that the plants' roots have reestablished adequately.

GROWING IN RAISED BEDS

Raised beds are a way to build up the soil in an in-ground garden in order to avoid poorly draining or otherwise poor-quality soil. You may choose to build up the soil only in certain spots.In this way, a garden can have areas of elevation where needed, and the materials used to create the raised area, such as bagged or bulk topsoil and/or compost, can be of a higher quality or different material than those that make up the rest of the garden. Or you can build raised beds for the entirety of your garden.

Ten Tastiest

GREEN GIANT

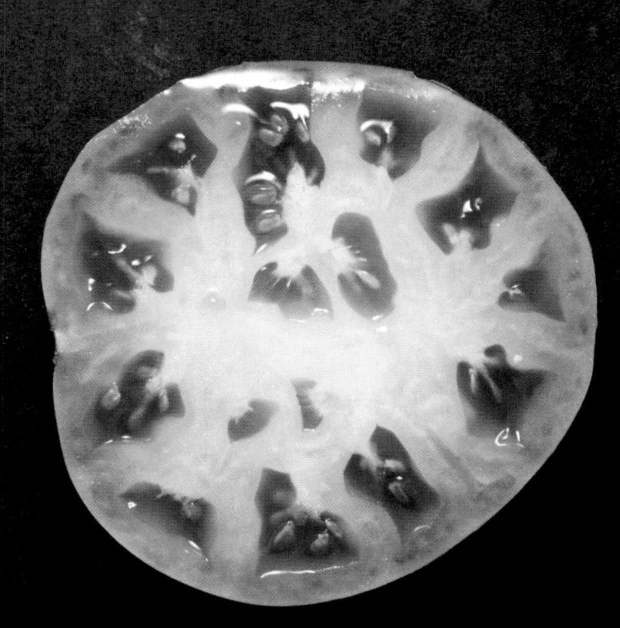

The flavor of **GREEN GIANT** took me by complete surprise at first bite. The texture, similar to Lillian's Yellow Heirloom and Brandywine, presents plenty of succulent flesh surrounding the many small seed cavities. The intensity of flavor nearly overwhelms, and it is perhaps most similar to Sun Gold in approaching perfection. As a class, large-fruited tomatoes that retain green flesh when ripe are uniformly delicious, but Green Giant edges out even Cherokee Green in its excellence.

You can raise the beds for your tomatoes with or without a confining structure. In the past, I've created raised rows using bags of soilless mix and planted the tomato plants deeply into the mix; excess water then drains from the raised bed area. Because my soil drained poorly, if I hadn't done this, my tomato plants would have sat with wet feet for prolonged periods after a rain, potentially drowning the plants.

A more typical raised bed garden is made up of well-built sets of squares or rectangles, typically from wood or concrete blocks, with wide aisles of gravel path or lawn that allow for complete access to all sides of the garden. These raised beds can be more work to create and more expensive to fill, but they are also well contained, neat, and more controllable. It's important not to crowd your plants as you plant in raised beds — as they mature, they could tangle into quite a jungle.

Pros:

- The beds can be attractive and give the impression of order and control.
- Crops can be compartmentalized into individual beds.
- The quality of the planting medium (soil or growing mix) can be controlled, at least early on.
- Raised beds warm up more quickly than in-ground beds. This in turn warms the root zone of the plant, accelerating growth.

Cons:

- There is less planting potential per area of ground.
- You have the added expense of building and filling the structure.
- Raised beds can potentially allow disease organisms to build up, which results in the need to replace the planting medium or solarize the soil.

GOING MOBILE — GROWING IN CONTAINERS

Because of the geography of your yard, you may find that there's no place to site a garden where it will receive enough sunlight to grow tomatoes. Fortunately, it's possible to have great success by bringing the plants to the sun — whether that's on a deck, patio, or driveway — simply by growing in containers. Even if you have enough sun, you might want to use containers if you have a buildup of disease in your soil or poor drainage. And growing in containers does not mean your options for tomato varieties will be more limited. Given sufficient container size and an appropriate plant support system, even indeterminate, large-fruited tomatoes can excel when grown in containers. My own experiences over the past decade (in which I've grown more and more of my tomatoes in containers) taught me that tomatoes grown in containers can equal those grown in the ground, in every way imaginable.

Pros:

- If you can grow it in the garden, with few exceptions, you can grow it just as well in containers.
- You can start with sterile containers each year and use fresh potting mix to avoid diseases.
- Container placement is flexible; wherever there's sun, there can be plants.
- The plants' roots are more easily warmed by the sun. The enhanced heating accelerates growth and development of the plants, providing earlier and significantly heavier yields.

Cons:

- The containers can be expensive (though this is often just a start-up cost, and there are plenty of inexpensive options), and the mix that fills the containers must be purchased, as well. (This is my most significant annual expense.)

Using creative staking and container gardening allows anyone with good sun exposure to have a productive garden. Every year my driveway becomes my main area for growing tomatoes, along with peppers and eggplant.

- Containers must be sanitized prior to reuse, to remove pathogens that could infect plants.
- Potting mix should not be used for more than one season, as diseases can transfer to the mix. (You can, however, reuse it as a planting base for other crops, such as lettuce and beans.)
- A container can become unstable when the plant is fully grown, if it's not properly staked.
- Vigilance is necessary to ensure sufficient moisture for the plants. (Drip irrigation solves this nicely. Without it, on very hot days, you may need to water mature plants twice per day.)

Containers

Anything that holds soil and drains through the bottom works fine for a planting container. I've used plastic pots in varying colors; the large pots that nurseries use for shrubs or trees are perfect. Grow bags, made of fabric or plastic, come in varying sizes and work wonderfully. They are durable, can be reused many times, and can withstand annual bleaching without loss of structural integrity; some even have handles for easier transport. Clay pots are generally fine but could be problematic in very warm areas, as the porous material is prone to greater moisture loss from evaporation.

VOLUME

I've found that large-fruited indeterminate tomatoes require containers with a volume of at least 10 gallons to perform as well as they would in a typical in-ground garden. Dwarf and determinate tomato varieties will do just fine in 5-gallon containers. Recently, I've been experimenting with ways to get enough fruit from indeterminate plants for evaluation and seed saving using 2-gallon containers; this requires some specific pruning and staking techniques and close monitoring to ensure adequate nutrition and moisture.

PLANTING MEDIUM

For my own containers, I combine one 2.5-cubic-foot bag of commercial soilless mix with one 25-pound bag of composted manure. An alternative that may be more cost-effective is to create your own soilless mix by combining sphagnum peat moss and composted softwood bark (in a 3:2 ratio), adding some perlite, perhaps a bit of a wetting agent and a trace of slow-release fertilizer, if desired, and composted manure. Screened compost can also be used. Avoid using garden soil to fill containers — it won't drain well in a container, and also carries the risk of introducing soilborne diseases.

CHALLENGES WITH CONTAINERS

Keep in mind the two major challenges with container growing: providing sufficient water and preventing the container from toppling over once the plants have grown large and become laden with fruit. Setting up drip irrigation solves the watering problem but has its own issues, such as cost, effort to erect, and suitability for the location of the garden (for example, since my container garden is our driveway, setting up a large drip system in a location that is necessarily temporary is more work than I'd like to undertake). Though it takes daily monitoring, hand-watering containers offers frequent opportunities to monitor progress, identify developing issues, and ensure ripe fruit is harvested when it's ready. Self-watering containers are a good, effective option but are more costly, and the impact of this expense grows with the size of your garden. Methods that I employ to keep container-grown tomato plants upright are described on page 88.

SUPPORT AND MAINTENANCE

Congratulations! You've done a lot of hard work to get to this point. You've done lots of digging, testing, mixing, and lugging, and it will all pay off in a few months when the harvests begin. With the tomatoes planted, cutworm collars in place, and plants mulched and watered, it's time to consider the ways to physically support them.

Types of Stakes

The vast majority of tomato varieties benefit from being staked. For indeterminate tomatoes, the taller the stake, the better; such varieties will easily reach 8 feet or more by the end of a growing season. Dwarf and determinate tomatoes, which can grow up to 4 feet, require a stake of equal height to keep them from sprawling.

Good tomato stakes can be made of any strong material — most are wood, plastic-coated metal, or bamboo. Whatever material you choose, a stake must be durable enough to withstand being hammered as deep into the soil as possible. It must also stay upright when burdened with a fully fruiting, tall plant experiencing the gusts of a summer thunderstorm.

Securing a stake close to the base of the seedling helps to ensure that the plant won't bend at the bottom once its top becomes heavy with fruit. Since tomato plants grow quickly, tying them to the stakes is often a weekly activity.

GROWING TOMATOES IN STRAW BALES

A local gardening friend recently shared with me his fondness for growing tomatoes in straw bales — basically, a different version of container gardening. There are many different ways to go about the process, including variations on when and how to wet the bales and what kind of fertilizers to use for pre-treating the bales. Here are a few of the basics:

1. Locate and purchase some straw bales and set them in an appropriately sunny location, whether in the garden or on a patio or driveway.
2. For the next three days, wet the bales well. Then, for the next week, wet the bales daily with a mixture of water and bonemeal or fish emulsion. This will help the bales start decomposing, which will create warmth.
3. At the end of the week, stick your hand into the middle of a bale. If it doesn't feel warm, it's time to plant, otherwise wait a few more days.
4. At 2-foot intervals, create holes 4 inches deep in the bales. Set a transplant in each hole, cover the root ball with sufficient compost or soilless mix, and create a slight depression or well around the stem; water well.
5. Water the bales regularly to ensure that they stay uniformly moist, and feed each plant every two weeks with dilute fish emulsion or compost tea.

Tie, and Tie Again

Tomatoes are referred to as vining plants, but they don't climb by automatically attaching themselves as they grow, as ivy or morning glories do. They need to be continually secured to the vertical support using ties made of relatively soft material, such as sisal twine. Make the initial tie of the vine to the stake at 6 inches above the soil. Rather than tightly binding it to the post, tie it loosely to allow for growth through the season. I use the twine to tie the plant to the main stake every 6 to 12 inches, depending on the behavior of the individual plant. Some grow more rapidly and rampantly than others, and some varieties seem especially floppy and weak (particularly many of the paste or heart-shaped tomatoes). I often use multiple stakes for a single variety if I choose to prune it minimally or not at all. Sun Gold is such a rampantly growing, heavily yielding type, and is so popular in our family, that it receives the multiple-stake treatment every year.

Closely securing the main growing stem to the stake prevents the naturally weak tomato plant from flopping over. If the tying job is ignored for too long, vines can easily snap with the weight of developing tomatoes.

Staking Containers

If you grow tomatoes in containers, providing adequate support becomes a matter of location. Of course, you can provide a stake of suitable height in the container itself, and it will be effective as long as the plant is young and not yet heavily laden with foliage, branches, and developing tomatoes. Once it grows in, however, it will need additional support.

Appropriate placement is an easy solution to the issue. If you position the tomato containers at a driveway or patio's edge where it meets the lawn, you can pound stakes of appropriate lengths into the ground and push the pots up against the stakes, providing all of the support needed for vertical growth. With this approach, I find it helpful to settle the transplant near the edge of the pot, which minimizes breakage near the bottom of the plant as it is stretched to meet

LET THEM SPRAWL?

Nothing is easier than planting a tomato and letting it roam free. Tomatoes grown this way are not pruned, and potential yields are enormous. With this method, it's important to consider all of the potential issues that would negate the advantages of ease and higher yield:

- **Most soil contains pathogens that attack tomato plants and fruit, so sprawling plants must sit on a thick layer of mulch that will ensure that soil does not come in contact with any part of the plant.**
- **Sprawling plants can swallow up many square feet of gardening space.**
- **Once the vines extend outward, it will be difficult to get to the ripe fruit in the center of a plant without stepping on and damaging parts of the plant.**
- **Slugs will cause significant damage to fruits that are in contact with the ground.**

the stake. If you have plants growing in larger pots (10-gallon minimum), you can use the large containers to anchor plants in adjacent smaller containers. In my driveway garden, I have 10- or 15-gallon pots containing indeterminate tomatoes all along the perimeter. In front of each of those pots I place a 5-gallon pot or grow bag with a dwarf tomato variety, placing the stake for supporting the dwarf into the larger pot. This prevents the dwarf plant from toppling over once it becomes mature and laden with fruit.

If your containers are on a deck and the pots are pushed up against a railing with vertical slats, try lashing a taller stake to a slat. This provides a perfect way to grow your tomatoes vertically and avoid tipping the pot.

Caging Tomatoes

Growing tomatoes in cages is a way to maximize yield (because they are typically not pruned) and minimize effort (once the cage is in place). Most often, tomato cages are fashioned from concrete reinforcing wire, manipulated to form a ring of varying height and diameter most appropriate for the garden size. Tomato cages are generally 4 to 5 feet tall and 3 feet in diameter and must be secured at two points by posts of similar height to prevent the plant from toppling over at maturity. The wide openings in concrete reinforcing wire make it easy to reach in and pick the ripe tomatoes.

For this technique, the tomato seedlings should be spaced so that a foot or more of space remains between the cages. To plant, set the

Training tomatoes vertically with stakes and cages allows for closer planting. Mulching with shredded leaves and grass clippings around the plant and between rows helps prevent soil from splashing onto the lower foliage, thus minimizing the spread of soilborne diseases.

seedling into the ground and then center the cage over the seedling (it is helpful to provide a single stake of the height of the tomato cage to ensure initial upright growth). Secure the placed cage at two opposite sides by stakes or posts, and tie the tomato to the center stake as it grows. The plant, which is most typically grown unpruned, will quickly fill the cage, and if the soil is rich and full of nutrients, the potential yields of caged tomatoes are enormous. The main concerns with caging tomatoes are cost of the cages, work needed to set up the cages, and finding an adequately large area to store the cages during the off-season.

SMALLER-SCALE CAGES

Big-box hardware stores typically sell tomato cages that are less than 5 feet tall and look like a narrowing cone, with prongs at the bottom and three rings of decreasing diameter attached to the outer prongs. Many other variations on tomato cages exist as well, including collapsible examples that end up as squares. These smaller-scale cages are quite inexpensive, but of limited and specific use. Those who plan to grow indeterminate tomato varieties (most large-fruited, colorful heirlooms fit this category) will be disappointed in such tomato cages, as the plants will quickly outgrow them. However, they are perfect for determinate and dwarf varieties, which grow to 4 feet tall or less. The short, cone-shaped tomato cages are also well suited for peppers, eggplants, and tall flowers.

VARIATIONS ON STAKING/CAGING

Most tomato varieties are indeterminate, meaning infinitely growing (until killed by frost or disease). They are often referred to as "vining," but unlike most vines, they don't wind or adhere themselves to adjacent structures; they must be supported and secured. Since vertical support removes most of the plant and all of the tomatoes from contact with the soil, it minimizes a mode of disease entry into the plant (as long as the area around the plant is well mulched). Once the decision has been made to stake the tomatoes, you will find that variations on staking tomatoes are infinite. It really comes down to the creativity and imagination of the gardener, available space, and budget.

Overhead Suspension

If a garden is large and will be maintained in one location for many years, a good way to avoid the cost and labor of pounding in many stakes is to grow tomatoes up a strong line or chain, suspended from above. This method requires the placement of very strong, tall posts at each end of a row of the garden, perhaps even cementing them into place. A chain is suspended between the two posts at the top, and at intervals where the tomatoes will be planted, strong twine droplines are tied to the overhead line and

The overhead piping in this structure (created by my local tomato-growing friend Richard Baldwin) ensures good vertical growth. Heavy string is suspended from the pipes into the containers below, and the tomato vines are secured to the string as they grow.

suspended down to the ground. Once the tomatoes start to grow, they are tied to the dropline in the same way as if it were a stake. If the gardener is very vigilant, the tips of the vines can be wound around the twine as they grow upward; however, the rapid growth of many plants will invariably lead to the need to do some remedial tying. This method does not allow for easy crop rotation, so disease buildup in the soil could be an issue that becomes more serious each year.

Teepees

Another option is to create teepees from three or four tall, slender poles, spaced at the bottom but meeting and joining at the top. A tomato seedling is planted at the base of each teepee pole and secured to the pole as it grows upward and outward.

Tomatoes grown this way often have several advantages. Compared to tomatoes grown in a single row, they have less of a chance to develop sunscald, simply because some of the fruit will be in the center of the structure as they form and mature. The structure itself is more stable than vertically staked plants in rows, and less likely to topple in the wind. It also creates a unique visual focus in the garden that could be quite dramatic.

The Florida Weave

The Florida weave technique, also known as the cat's cradle, is an effective, easy, and relatively inexpensive way to support indeterminate varieties of tomatoes grown in rows. The key is to provide robust support (able to withstand heavy plants and the winds that come with summer thunderstorms) at the end of each row.

Mark out your tomato rows, spacing a minimum of 3 feet between rows. At the ends of each row, hammer sturdy poles, such as steel T-posts used for fencing (preferably at least 6 feet tall), deeply into the soil. Plant the tomato seedlings down the row, using an optimal spacing of 3 feet between plants if possible, and hammer additional poles into the row after every two plants.

THE JAPANESE RING

One interesting variation on caging is called the Japanese ring. The technique calls for tomato seedlings to be planted around the outside of a wire cage that is filled with nutrient-rich organic material. This serves as the "feeder pile" — a reservoir of nutrients and moisture into which the roots of the tomato plants will grow.

To create a Japanese ring, bend a 9½-foot length of 6-foot-tall concrete reinforcing wire into a circle and support it with four metal garden stakes driven into the soil. Into the center of the cage, heap a 2-foot layer of rich planting medium (such as a mixture of compost and fertile topsoil). Create a depression in the center of the pile to enhance water retention.

Spread a thick layer of the planting medium around the outside of the cage, and plant tomatoes equidistantly around it (4 plants per cage). Water the pile as needed, and add a balanced fertilizer or fresh application of compost every 2 or 3 weeks, if necessary. Tie the plants to the cage as they grow. The plants will eventually grow up the cage and into the center, and they will load up heavily with fruit.

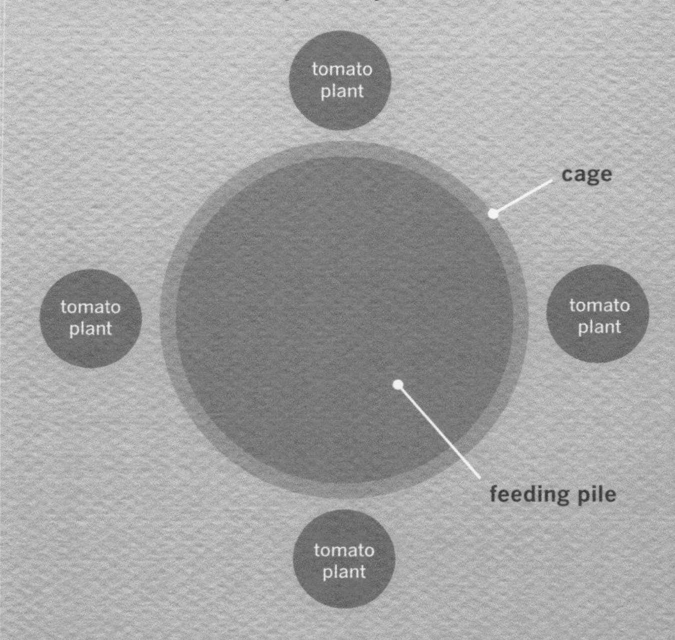

Secure sisal twine to one of the end poles at about 10 inches above the soil line, and run the line along one side of all the plants, looping it around the internal poles. At the end of the row, loop the twine around the other end pole and bring it back down the row so that the twine provides support for the plants on the other side, again looping back around the internal poles and then securing it to the original point.

Repeat this process as the plants grow (every 10 to 12 inches). Plants grown this way may benefit from a moderate level of sucker removal in order to prevent sucker growth from cascading downward and creating an impenetrable tangle.

TO PRUNE OR NOT TO PRUNE?

Aside from questions on my favorite tomato varieties, I am most often asked about pruning and suckering of tomato plants. Suckers, or side shoots, are additional fruiting stems that emerge all along the plant at the junction of the main stem and leaf stems. If the leaf stem emerges from the main stem at an approximate 90-degree angle, the sucker will be at a 45-degree angle. The reason that indeterminate tomato plants grow out of control so quickly is that the suckers themselves go on to produce suckers, and a plant can become densely complex by midseason.

Should You Snip Your Suckers?

An indeterminate tomato plant extends its main growing stem indefinitely and generates suckers at each point of a leaf stem attachment. Essentially, suckers are the tomato plant's way of passing on its genetic heritage by providing a fail-safe mechanism for producing as many seeds as possible; the more branching, the more flowers, which lead to more tomatoes, which lead to more seeds, which to a plant means future survival. Contrary to pervasive urban legends, they do not sap energy from the main tomato plant, and allowing them to develop does not delay fruiting or ripening of any tomatoes from the main stem. Removal of suckers has implications for eventual yields and for how the plant needs to be maintained, so a detailed discussion on retention or removal of suckers is an important one to have.

Suckers, or side shoots, develop at every node between the stem and leaves up the vines of indeterminate tomato varieties. Some gardeners remove them, but leaving them in place will not negatively affect the plant.

WHAT SUCKERS PROVIDE

Each sucker allowed to grow provides additional flower clusters, and hence there are additional chances for fruit set. Here's an example to illustrate this point. Picture a tomato plant that has all of its suckers removed, tied to an 8-foot stake. A blossom cluster is produced at 8- to 12-inch intervals, starting at 2 feet from the soil line. During the season, the majority of the flower clusters open at times when the temperature and/or humidity is not suitable for pollination, leading to blossom drop. As a result, only a handful of fruit is produced on the 8-foot-tall

Ten Tastiest

SUN GOLD

The first time most people pop a **SUN GOLD CHERRY TOMATO** into their mouth and crunch down, the look on their faces is priceless. Sun Gold is unique in the range of flavors that it exhibits. Picked pale orange, there is fullness and complexity with a nice snappy bite. At medium orange, sweetness and tartness play out in a delicious dance of intensity. At maximum ripeness, they are like candy — nearly overwhelmingly sweet. In my experience, nothing in the tomato world tastes like Sun Gold. Because of this, it is the only hybrid that finds a place in my garden every year; no open-pollinated alternative exists.

USING SUCKERS TO EXTEND THE SEASON

An easy way for gardeners in warm climates to extend the harvest is to stagger the planting of the tomato seedlings. This can be done by starting a second round of seeds; the easier solution, however, is to simply root some of the suckers that the tomato plants are already producing so enthusiastically. Suckers root very quickly and easily, producing a clone of the plant they came from. Similarly, you can root the section removed when you top a plant, or even a young plant taken down by a cutworm (if you get to it before it wilts in the sun).

To do this, cut a 6-inch-long sucker from one of your healthy plants and put it in a glass of water, or push it gently into a pot of well-moistened potting mix. Some gardeners find that suckers will root directly in the garden if placed in a shaded area. It is important to keep the rooting suckers out of direct sun. Typically, a sucker will root and begin to show renewed growth in a few weeks at most. Often, the sucker in the moist soil will wilt just a bit until new roots begin to grow.

For a sucker rooting in a glass of water: once it has a nice set of new roots, pot it up in fresh potting mix and allow it to adjust in a shaded location for a few weeks before setting it into its final location.

plant, with no mechanism available for producing additional flowers. If just one sucker would have been maintained, the number of flower clusters would have doubled, and it is highly likely that flowers on that additional growing shoot would have opened under more suitable conditions, thus significantly increasing the yield of the plant. Following this logic, each sucker or side shoot that's allowed to develop will significantly increase the yield potential of the plant (given a long-enough season).

Suckers that are allowed to grow also provide additional foliage cover. In climates where the searing sun beats down on the exposed developing tomatoes, sunscald is a definite risk (see page 200). Since direct sun is not needed to promote ripening, it is far more preferable to have the fruiting cluster shaded by foliage.

WHEN TO SNIP SUCKERS

An indeterminate tomato plant can grow out of control in a hurry due to the formation of suckers, and then suckers from those suckers. Disciplined removal of suckers in order to provide a plant with a finite number of fruiting stems or branches leads to more control over growth and a far easier support task. In a plot with a very crowded layout, pruning to one or two stems will allow for more air circulation between plants and a way to maintain order in what could otherwise become a tangled mass of vines.

Topping Plants

In addition to snipping suckers, skillful "topping" of fruiting tomato branches and stems at particular heights is another way of maintaining control over a plant. Since more flowers form than will pollinate and ripen before the end of the season, topping also ensures that a plant doesn't put energy into developing tomatoes that would never get a chance to properly ripen. To do so: pick a plant height equal to the length of the supporting stake, and, with clean shears or fingers, pinch the stem just above a leaf stem that sits close to the final flower cluster from which you desire fruit set. Topping is an excellent way to prevent plants from becoming so top heavy that they topple in storms or develop kinks in branches, which can lead to disease and death of the plant above the kink.

When *Not* to Prune

Determinate tomato varieties are unusual in that they grow to a genetically predetermined height and width and then produce flowers at the end of the flowering branches, thus limiting outward growth. In a sense, they are self-pruning, and any removal of suckers will reduce the eventual yield of the plant.

Dwarf varieties, though they grow and behave like very slowly vertically stretching indeterminate types, don't require pruning to prevent uncontrolled growth and sprawl. A few suckers can be removed in order to open up the center of the plant — the foliage of dwarf tomatoes often grows densely, and removing some of it will encourage good air circulation, which is especially helpful in hot, humid gardens where foliar diseases thrive. Keep in mind, though, that pruning suckers from dwarf tomatoes will reduce yields.

WATERING

Adequate, even watering is essential for healthy plants. Tomatoes grown in soil, in a traditional garden bed, often receive sufficient rain throughout the growing season to render the need for deep watering infrequent, especially when the soil contains a lot of clay. Well-mulched plants may show some wilting during the hottest part of the day, but in the evening evapotranspiration decreases and the plants bounce right back. This, combined with the ever-deepening water-seeking roots, remedy the temporary water shortage. If it is dry, a weekly deep watering will suffice; a good workaround is the use of drip irrigation, which provides a constant water source for the plants.

Container-grown plants require much more frequent watering because they have less soil and less available water per root area for each plant. When the seedlings are very young, weekly watering works well, but once growth takes off, in the absence of substantial rain, daily watering (each morning) will be necessary. Fully grown plants during a heat spell should receive morning and late afternoon watering. My method is simple: with no nozzle and hose running full, 5-gallon pots receive a count to three, and 10-gallon pots a count to five, resulting in a bit of excess water eventually draining out of the bottom of the pot.

Don't Wet the Foliage

The mantra of proper watering is: water from the bottom, and never wet the foliage. When you think about it, this can be controlled only part of the time; clearly a heavy rainstorm comes in from above and saturates the tomato foliage. Proliferation of foliar diseases is always a follow-on possibility, particularly if disease

Water at the base of the plant to keep the upper foliage as dry as possible; use a thick mulch to keep the planting medium from splashing onto the lower foliage.

agents are already present, the plants have a prolonged period of wet leaves, and it is followed by high heat and humidity. Nevertheless, by being mindful and watering the soil instead of the foliage, you can at least reduce *your* role in the proliferation of foliar diseases.

One of the most important things you can do before you go away for vacation is to make sure someone whom you trust, and who understands your garden's needs, is on hand to provide the necessary watering regimen while you are gone.

FERTILIZING

Rich garden soil that has adequate nitrogen, phosphorus, and potassium (information that can be gained by doing a soil test) will grow great tomatoes. The ideal ratio of these three key nutrients is 1:4:2 (highest in phosphorus). Too much nitrogen promotes foliar growth at the expense of fruit set. There are specific fertilizers for tomatoes; however, use of *any* fertilizer should be based on need (what the early soil test tells you, what your past experience has been, and plant performance during the season). If you produce your own compost, working it into the soil around your plants a few times each season may be all that you need. Use of fertilizers tends to be inexact, since there are so many variables associated with your own particular location.

In-Ground versus Containers

Where you're growing your plants — in the garden or in a container — will determine how often you need to fertilize. Nutrients are retained longer near the root zone of garden-grown plants. The frequent watering of containers means that nutrients wash out the bottom more quickly, which calls for more frequent fertilizing. No matter how they're grown, the foliage color is a good indication of the need for a boost; the rich, dark green starts to fade to a paler hue if a plant needs some nutritional attention.

Organic or Synthetic?

Once the plants are up and growing vigorously, the fertilizing philosophy of the individual gardener comes into play, especially when deciding between chemical (such as Miracle-Gro or Vigoro brand products) or organic (such as Tomato-tone or fish emulsion) methods. With so many options of products and materials for both approaches, it is well worth the time for all tomato growers to do their own research, including trial and error, though my approach can be used as a starting point.

I've tried both chemical and organic fertilization approaches with good results throughout my gardening years, but I've settled most recently on using a high-quality, slow-release, balanced fertilizer for vegetables, such as Osmocote or Vigoro tomato food, applied every few weeks, because it fits my current emphasis on container gardening. I increase the frequency of feeding to weekly once my container-grown plants put on a heavy set of developing tomatoes.

A LITTLE WILTING WON'T HURT

Keep in mind that a bit of wilting isn't a problem. Wilted plants recover in a hurry, and even well-watered plants will show some foliage wilting when the sun is overhead and the temperatures hover in the 90s or higher; this is the plants' way of conserving moisture. Real vigilance for severe wilting during hot spells is needed if the plants carry heavy sets of green fruit; the uneven watering experienced by a plant that dries to wilting, followed by a gush of water, often leads to blossom end rot. Mulching helps to retain moisture. Drip irrigation is a great solution, and using planting mix that has water-retentive crystals, though costly, will provide help as well.

TOMATOES WITH HISTORY

I love genealogy and am quite nostalgic; that trait has applications in the pursuit of old tomato varieties, as we try to find out what our grandparents, or their parents, may have grown in their gardens. The best tool for this type of sleuthing is old seed catalogs. The catalogs provide the names of vegetables in American gardens as far back as the mid-1800s.

With information from old catalogs in hand, it comes down to knowing where to look for the varieties, if indeed they still exist. Sadly, many are now extinct, though it is surprising how many of the old types are still available to be grown in our gardens. One useful resource that I started using in the early 1990s is the Germplasm Resources Information Network (GRIN), a portal that allows a search of the plant materials in the National Genetic Resources Program (NGRP), within the research service of the USDA. With the names of many old tomatoes in hand, I was delighted to find that many varieties thought to be long gone resided in the NGRP, and I requested and received samples from dozens and dozens of treasured old tomato varieties.

Ferris Wheel

Occasionally, I order a tomato just because of its fanciful name. One such variety, Ferris Wheel, made its way into the NGRP collection in 1943 according to the information provided on GRIN. I was in no rush to grow it until I was looking through an old catalog from the Salzer Seed Company and I noticed that Ferris Wheel was one of their tomatoes, with a release date between 1894 and 1898. The Salzer description was quite over the top in superlatives.

After realizing that Ferris Wheel was an authentically very old variety that likely hadn't been grown for many years, I gave it a try, in a 15-gallon pot in my driveway in 2001. The plant had the characteristic somewhat-wispy foliage and open habit of many of the large pink beefsteak types. The fruit were in the size range of 1 pound or more, were pink with pronounced green shoulders, and had some ribbing and irregular shapes. The flavor was just remarkable — full and sweet, complex and delicious. Ferris Wheel became an instant favorite, and I grow it often. My guess is that it is quite similar to an old variety released by the Henderson Seed Company in 1891 called Ponderosa. In fact, Ferris Wheel could indeed be a "selection" that the Salzer company made from Ponderosa, perhaps from a plant showing a distinct improvement in a particular characteristic. It certainly is a winner, and one can only wonder why it disappeared from gardening awareness for so long.

SALZER'S FERRIS WHEEL TOMATO.

Tomatoes of this monster, this leviathan, this giant, have been **grown** weighing **5 lbs.**, and measuring almost **2 feet in circumference.** We studied for an appropriate name and finally concluded to name it Ferris Wheel Tomato, in honor of one of the greatest inventions of the age. This Tomato possesses every good quality to be found in a Tomato. The vines are strong and vigorous and easily bear their enormous weight of fruit, and when we come to giant size, weight, solidity, no other Tomato begins to approach it.

I. J. Schaeffer, New Mahoning, Pa., says: **"Salzer's Ferris Wheel Tomatoes are immense. I raised them weighing 3¾ lbs. I sold them at 40c a basket: other farmers sold their kinds at 20c a basket. Salzer's Tomatoes brought 100 per cent more.**

W. W. Locher, New Castle, Pa., says: "For the past eight years I have been a Tomato specialist. I have tried every known variety, and I want to say that I unhesitatingly and unqualifiedly pronounce **Salzer's Ferris Wheel** in all particulars the finest Tomato I have ever grown or tested. It is the most prolific, the finest flavored, the most solid and beautiful of them all. It is a very early, a fine medium and a splendid late Tomato. It's an ever-bearer."

Price of Ferris Wheel: Pkg., 20c; 3 for 50c; ½ oz., 60c; oz., $1.00; ¼ lb., $3.50; lb., $10.00.

Price of Ferris Wheel; Pkg., 20c; ½ oz., 60c; oz., $1.00; ¼ lb., $3.50.

ABRAHAM LINCOLN

Surrounding this historic variety are a few delicious mysteries. The first time I heard of it was while reading *The Total Tomato* by Fred DuBose. Mr. DuBose had a very helpful listing of recommended tomatoes ranked by stars (from one to three), and Abraham Lincoln was among his very few three-star tomatoes. Though it is open pollinated (non-hybrid), it is considered a commercial, rather than family, heirloom. It was the cover tomato — the feature release — of the Buckbee Seed Company catalog in 1923. In that catalog, Abraham Lincoln is described as being very large, nearly round, red, extremely productive, and delicious. The variety was sometimes described (particularly in the entry in Mr. DuBose's book) as having unique dark-toned, bronze-colored foliage that stood out in a garden.

For a variety that is reasonably recent and has been continually available for gardeners to grow through seed catalogs, obtaining seed that delivers what the old catalogs describe continues to be quite a challenge. When I grew out Abraham Lincoln tomato seed from the Shumway Seed Company back in 1987, the plants were green, semi-determinate, and less than 5 feet tall, and they produced tomatoes that were only 3 or 4 ounces — not at all what was described in other references.

A few years later, I grew out seedlings from another supplier, and though it didn't exhibit the unique bronze or purplish cast noted in the book entry, it was certainly indeterminate, and the fruit were large, closer to a pound. The flavor was quite delicious. Of course, there was no way to confirm if this was Abraham Lincoln as released by the Buckbee Company in 1923. It was, however, closer in description to the authentic variety than any I had grown to that point. And so I grow it still for its flavor, productivity, and history. It very well could have been a variety that my grandfather grew during his gardening years. And that possibility makes me quite happy to be growing it.

Big Boy

I include Big Boy here and in my garden purely as a testament to its historical importance. Though not strictly the first hybrid tomato, Big Boy was certainly the first wildly popular, revolutionary hybrid tomato. Released in 1949, it was part of Americana following the end of World War II that boosted morale and, in a way, defined the Victory Garden. Big Boy had an impact on tomato size, yield, future breeding efforts, and seed company focus going forward.

Prior to Big Boy, seed companies, the USDA, and university agricultural programs involved in breeding new tomato varieties worked toward releasing stable, open-pollinated types that could be replicated from saved seed. Big Boy changed all that. Creating hybrids is more labor intensive, so seed cost rose. Seeds couldn't be saved, so customers had to return annually for seeds of hybrid varieties. Many of the new hybrids were also bred to resist disease. Big Boy also whetted the gardeners' appetites for large, smooth, round, red tomatoes.

All of these factors pushed older, heirloom varieties, with their odd colors and irregularities, further into the background as the public rushed to grow the latest and greatest hybrids. Big Boy was followed by Better, Ultra, and various other designations of superiority, and Girls joined the Boys as well.

Burpee's Hybrid Tomatoes

ALL CREATED ON BURPEE'S FORDHOOK FARMS

Unsurpassed for quality, uniformity and yield. Every seed from a hand pollinated fruit.

1131 BURPEE'S BIG BOY

Shown in color on back cover

78 days. Largest of our true, first generation (F_1) hybrid tomatoes; a splendid companion to other Burpee Hybrids.

Well grown fruits average 10 ozs. with some weighing 1 lb. or more. Unlike other large fruited varieties, Big Boy tomatoes are perfectly smooth, deep globe to deep oblate in shape and very firm; scarlet-red skin, thick walls, with bright red meaty flesh of fine flavor and excellent quality. At its peak of performance in midseason, when many standard varieties decline in yield and size of fruit. An ideal hybrid for the home garden and market grower, especially roadside sales.

Plants are very large, extremely vigorous, semi-upright to spreading in habit; moderately dense dark green foliage protects the fruits from sun-scald.

The picture at left gives you some idea of the height of a staked plant, the size of the tomatoes and the quantity of fruit produced. Because of the extra strong, thick, sturdy, strong stemmed growth, be sure to provide tall, strong poles, preferably rough, to better hold string or other material used to tie up the plants. **Pkt.** (30 seeds) **50¢; 2 pkts. 90¢; 3 pkts. $1.25; 5 pkts. $1.90**

WHY HYBRID VEGETABLES?

It is a well-known fact that true hybrids grow faster than open-pollinated varieties, show more vigor, produce greater yields and often, over a longer period of time, bear a more uniform product. And, very often, it is possible to incorporate into the hybrids resistance to a number of diseases and insects. Many hybrids not only combine the best features of their parents but even far surpass the best that either parent has to offer.

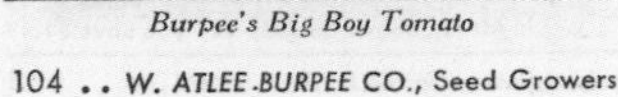

Burpee's Big Boy Tomato

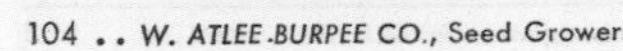

104 .. W. ATLEE BURPEE CO., Seed Growers

Ten Tastiest

Lucky Cross

Belying its regal heritage, **LUCKY CROSS** is a flavor king among the large yellow-red bicolored tomatoes. As an offspring of Brandywine, Lucky Cross shows the same sort of big tomato flavor and perfect balance, rendering it unique among varieties in this color class, which typically have a very sweet flavor personality that borders on overly mild.

PLANTING CONTAINERS, STEP BY STEP

1. Prior to planting, immerse each container in a dilute bleach solution and scrub well. Allow the container to drip-dry in the sun.

2. Fill each pot with dry soilless mix, leaving 3 to 6 inches between the top of the mix and the upper edge of the pot, depending on the pot size (I fill the small 1-gallon containers that I use for my hot pepper research nearly to the top, but I leave about 6 inches of space for the 15-gallon pots used for indeterminate tomatoes).

3. Scoop out a deep impression and settle the seedling into the mix, firming it around the stem of the plant. Plant the seedling deep in the container, burying as much of the stem as you can.

4. Water the pot well from the top, ensuring that the foliage doesn't become wet, and apply a layer of organic mulch to conserve moisture. Position the pot next to a stake that has been driven into the ground or into another container, and loosely tie the plant to the stake. Continue to tie the plant as it grows up the stake.

Harvest CELEBRATION

ALL OF THE HARD WORK IS NOW behind you (so you think, anyway!). The journey that began many months ago is beginning the lengthy (we hope) payoff period. Harvest time often begins utterly unexpectedly with a flash of color deep among the dense lower foliage of one of the tomato plants. One day, the tomatoes are hard and green, the dinner plates yet to be decorated with the anticipated harvest, the stomach rumbling with anticipation. And then, there it is: the first ripe tomato, just waiting for you to pick it, devour it, relish the flavor of summer, and become immersed in the nostalgia of so many tomatoes tasted throughout your gardening years.

DETERMINING WHEN TO PICK

There are different schools of thought on when to pick a ripening, or ripe, tomato. The decision depends upon several factors: the variety, the weather (temperature as well as precipitation or irrigation), the presence of marauding critters, and the status of your edible ripe tomato supply and its intended uses. It also depends on your knowledge of the variety and expected color, whether the variety is true to type, and where you are in the season (early, mid, or late). Trial and error will certainly be involved, and experimenting with a tomato you are trying for the first time can be very informative.

Generally speaking, when tomatoes are ripe for eating, they're also ripe for seed saving. Good, viable seeds can be saved from tomatoes that are a bit short of fully ripe (when they are blushing, but still showing some green), though germination could be a bit low. Seeds saved from overripe tomatoes will be fine.

Peak Flavor

Of course, the flavor of a particular tomato will be at its peak when it's perfectly ripe. Each variety, at peak ripeness, will also possess a characteristic texture. As a rule, less-ripe tomatoes have less sugar and could exhibit a more acidic, tart taste. The texture is likely to be firmer, perhaps even with a definite hardness or crunchiness. Tomatoes that ripen a bit beyond the peak state will become sweeter and less tart, which could result in unpleasant blandness. The texture in this case may become mushy or soft. Sometimes, an overripe tomato will begin to show nearly rotting "off" flavors that make the tasting experience very unpleasant indeed. It is wise to stay away from preserving or processing tomatoes that are overripe to the point of near collapse, as the off-flavors will taint your efforts.

Texture

Another aspect of ripeness is a softening of the flesh: give the tomato a gentle squeeze. Though a yielding of the flesh is a good general guideline, some varieties, particularly some of the more modern hybrids, are bred to be particularly firm, and even fully colored specimens will feel surprisingly firm.

Aroma

I've watched many shoppers at farmers' markets pick up tomatoes and smell them. Personally, I get little to no aroma from tomatoes (they are very different from slip melons, such as cantaloupes, in this regard); often, a strong smell associated with a ripe tomato is indicative of impending spoilage. Of course, each person "smells" differently, so others will have quite a different opinion or experience than mine.

Ripeness by Color

Tomato colors are determined by the genes of each variety. The relationship between a particular tomato variety's color potential and ripeness level is something best learned through experience. The particular mixture of carotenoids in the tomato flesh, including lycopene (yellow, orange, and red, as well as red and yellow bicolored types), and chlorophyll (green, which also contributes to the so-called "black" tomatoes), when overlaid with clear, yellow, or striped skin, determines the apparent ripe color of the tomato.

An important thing to remember is that the temperature tomatoes experience when ripening can significantly affect the color. Fruits that are exposed to midsummer sun may develop the pale and withered coloring indicative of sunscald. A red tomato that ripens at temperatures above 85°F will not develop a rich, deep pigment and will appear a washed-out pink or even yellow.

Most gardeners are familiar with the color journey of red tomatoes on their way to ripeness. But when growing tomatoes of other colors for the first time, it can be confusing for the gardener to know the right time to pick. Many a customer has told me, "I didn't know when to pick the tomatoes because they never turned red, and I think they just rotted on the vine." That is a shame, because many of the

unusually colored tomatoes are among the very best in flavor.

It is also worth noting that tomatoes are edible along a continuum of apparent ripeness, from just past blush (fully colored, though perhaps paler than the fully ripened specimen) to as deep a color as the tomato reaches prior to the onset of rotting. When you throw in the factor of the uniqueness of each variety in terms of optimum ripeness, as well as variations from season to season, it becomes a matter of tasting to decide what you like, and when you will like it best with regard to picking time.

Red tomatoes — those that have yellow skins and deep pink or red flesh — go through degrees of color change on their way to full ripeness. The compound responsible for the red flesh is a carotenoid called lycopene. At the onset of color change, the tomato could be pale pink (indicative of the flesh color changing in advance of the skin color) or pale orange. As ripening continues, the fruit starts to take on the characteristic red-orange tone of the red tomato, with well-known examples being Better Boy, Roma, and Celebrity.

Early in the season the change from a faint tint of color to full, rich scarlet can take a frustratingly long time. Later, in the middle of the summer heat, when gradual ripening would be a welcomed occurrence (because of the glut of tomatoes becoming overripe on the kitchen counter), a faintly tinted tomato one day seems to become perfectly ripe in the blink of an eye.

Some varieties of red tomatoes will retain green coloring at the shoulders around the stem end. This genetic characteristic is often retained even when the lower majority of the tomato is fully ripe. Again, this is a case when familiarity with a particular variety is very helpful. I am sure many a tomato grower has mistakenly let some varieties rot on the vine while waiting for all of the green coloring to vanish.

Figuring out when pink tomatoes are ready to pick can be a bit tricky. Because the only difference between a pink and a red tomato is skin color (pink tomatoes have clear skin when ripe), not only is there great confusion even among experienced gardeners in determining whether a tomato is red or pink, but it is often impossible to make a final determination of color until the tomato is fully ripe. Red varieties that ripen during a hot spell may appear to be pink.

As pink tomatoes ripen, some varieties can, oddly, have a yellowish cast. Typically, they are pale, but occasionally dusky, shades of pink. Over time, they become a deep crimson pink. One thing to keep in mind if you are saving seed or cooking with a combination of red and pink tomatoes is that, once sliced, the tomatoes look essentially identical. Since red and pink tomatoes are basically identical in flesh color, layering cut slices of each type in a Caprese-type salad is not visually effective. Using whole pink and red cherry tomatoes in salads is a good way to highlight the color differences.

("BLACK" VARIETIES)

The various "black" tomatoes (with purple or brown hues) can be quite tricky to assess for ripeness, especially by those who are unfamiliar with them. Black varieties that retain green shoulders even when very ripe are a bit easier to distinguish; otherwise, they may appear from the outside to be simply pink or red varieties.

Tomatoes of all colors take on a deeper hue as they ripen on the counter, but I've found that this is particularly striking with the black varieties. It is also important to remember the correlation between temperature and pigment formation. It is, therefore, likely that people growing black varieties in different climates experience different depths of hue, leading to all sorts of interesting "discussions" (sometimes arguments) on Internet forums.

Red & Brown

JAPANESE TRIFELE BLACK

ROMA

ARKANSAS TRAVELER

FAVORITE

MORTGAGE LIFTER

CHEROKEE CHOCOLATE

Yellow & Orange

EGG YOLK

YELLOW WHITE

DWARF SWEET SUE

KELLOGG'S BREAKFAST

GOLDEN QUEEN

Since the appearance of and popularity boom for the black varieties is quite recent (beginning around 1990), we are all still learning about the optimum time to harvest and eat these uniquely colored varieties. If you're growing black varieties for the first time, be sure to observe the color changes closely as the tomatoes ripen, and pick fruit at various stages of ripeness, sampling them to get a sense of how they develop color and ripeness in your particular garden.

Assessing the ripe state of orange and yellow tomatoes is fairly easy once you know which color to expect. The continuum of hues is vast, from the palest canary yellow to deep gold, and it is more challenging — for me, anyway — to delineate yellow from orange with certain varieties. To further complicate things, some yellow tomatoes darken with age, and oranges and yellows possess different degrees of flavor at different colors. A perfect example of this is the popular, delicious hybrid cherry tomato Sun Gold. Since it is a tomato machine, you'll have plenty of opportunities to experiment with picking Sun Gold at different stages of ripeness. At pale orange with perhaps a faint greenish tint, Sun Gold has a very full flavor with a sharp acidic tang, which is quite delicious. As it passes to medium orange, the sweetness and complexity (some call it a tropical fruit element) comingle with the still-present tang, creating what many consider the perfect tomato-eating experience. Finally, Sun Gold will turn into a rich, deep orange, and the sweetness is nearly overwhelming. We've found uses for Sun Gold at each of these stages. However, care is needed, since this thin-skinned cherry tomato loves to crack just as it is turning from pale to medium orange, especially if the plant is heavily watered or enjoys a heavy rainstorm.

In general, the best guidance to offer is to experiment. Pick some orange or yellow tomatoes at various stages of ripening; taste some, and allow some to sit on the counter for further ripening and color development. From my experience, the evolution of flavor with increasing ripeness is present in all tomatoes, though at a lesser extent than with Sun Gold. There is so much to learn about how tomato flavors evolve with ripeness.

Let's look at a representative selection of larger-fruited (non-cherry) tomatoes along the yellow-to-orange range. The most pale-colored examples, such as Hugh's or Lillian's Yellow Heirloom, ripen to a very pale yellow and when extremely ripe often develop a pale pink blush at the blossom end. Still yellow, but of a richer, more buttery color, are Lemon Boy and Golden Queen. Once we move from yellow into orange, the presence of the pink blossom-end blush isn't typically seen, and the pale orange Persimmon and Yellow Brandywine are the color of a traditional pumpkin, whereas Kellogg's Breakfast reaches the deep hue of a navel orange. It isn't difficult to judge when any of these are ripe and ready to eat — the key is knowing what is to come, and not waiting for them to turn red!

BICOLORED. Quite a few of the yellow tomatoes have a tendency to develop a very pale pink blush at the bottom, and rarely, this will extend into the flesh as a bright crimson ring in the tomato core. Lillian's Yellow Heirloom is a perfect example of this. And taking things just a bit further finds us at the large, yellow tomatoes that are swirled and streaked with shades of pink and red. These tomatoes, the so-called bicolored beefsteaks, are, to me, the "peaches" of the tomato garden. Often the coloring evolves and grows as they sit on the counter. These bicolored beefsteaks are not at all uniform, and each tomato on the plant will develop different color patterns.

Hugh's

Ten Tastiest

LILLIAN'S YELLOW HEIRLOOM

Until I tasted **LILLIAN'S YELLOW HEIRLOOM**, my view on orange- or yellow-fleshed tomatoes was one of slight indifference; most of them were on the mild, approaching bland, side. How wrong I was. Lillian's Yellow Heirloom, though temperamental in terms of consistency and yield from year to year, can often elevate to the top ranks of tomato flavor. Possessing a superbly uniform, nearly creamy, juicy texture, it is a loud tomato that fills the senses, perfectly balanced between tart and sweet. It is an annual highlight on our tasting plates.

CUSTOM-COLORED GAZPACHO

Through the years, we've sampled hundreds of versions of this cool, fresh tomato treat. Many of them were truly delicious, but after this interpretation by chef Sarig Agassi (of Zely and Ritz restaurant in Raleigh, North Carolina), no other gazpacho can compare. It is outstanding, simple to make, and remarkable in the intensity of tomato flavor. Be sure to use only the best-flavored tomatoes, as they definitely take center stage.

This is a soup that really lets one play with hues — in terms of both color and flavor characteristics (full, gentle, sweet, tart). For a knockout bright yellow soup, try to find Hugh's, Lillian's Yellow Heirloom, Lemon Boy, or Azoychka. The large, meaty red-yellow bicolored types, such as Lucky Cross, Ruby Gold, or Pineapple, make for a gentle, sweet soup that exhibits sparkles of pink among the warm yellow tones. Be adventurous, be creative, and have fun with your tomato combinations, but beware when mixing green- and red-fleshed varieties: the resulting brownish color will not be nearly as appetizing as the flavor.

INGREDIENTS

- 12 large heirloom tomatoes
- 1 sweet onion or leek
- 1 cucumber
- 1 sweet pepper
- 2 tablespoons salt
- 2 teaspoons freshly ground black pepper
- ½ cup sherry vinegar
- 12 basil leaves
- Cherry tomatoes, halved, for garnish
- Extra virgin olive oil, for garnish

1. Dice the tomatoes, onion, cucumber, and pepper; combine in a large bowl.
2. Add the salt, pepper, and vinegar to the vegetable mixture, and let marinate in the refrigerator for 24 hours.
3. Add the basil and purée.
4. Serve slightly chilled, with some colorful halved cherry tomatoes and a swirl of extra virgin olive oil.

SERVES 6–8

COLOR-CODED SALSA FRESCA

We've all been to parties and sampled an innocuous-looking bowl of red salsa, only to find ourselves gasping, sneezing, or escaping to blow our runny nose because of the unexpected level of fire. This can be avoided by color-coding salsas by their level of heat. The recipes below telegraph the eating experience to the unsuspecting salsa victims: green is mild, amber is medium to intense, and red is hot.

FOR ALL THREE SALSAS

- 6 large tomatoes, chosen to impart the desired color for each salsa, finely diced
- 2 tablespoons olive oil
- 2 tablespoons red wine vinegar
- 1 tablespoon salt
- 6 scallions, finely sliced (including white and green parts)
- Freshly ground black pepper

GREEN SALSA

- 1 green jalapeño or serrano pepper, seeded and finely chopped

RECOMMENDED TOMATO VARIETIES:
Green Giant, Green Zebra, Cherokee Green, and Aunt Ruby's Green

YELLOW SALSA

- 2–3 jalapeños or serrano peppers (with seeds), finely chopped, or, better, a few yellow Santa Fe–type chiles

RECOMMENDED TOMATO VARIETIES:
Kellogg's Breakfast, Yellow Brandywine, Lemon Boy, and Lillian's Yellow Heirloom

RED SALSA

- 2 red ripe Thai chiles, finely chopped
- 1 red ripe habanero (with seeds), finely chopped (or more, depending upon how hot you wish to go)

RECOMMENDED TOMATO VARIETIES:
Brandywine, Andrew Rahart's Jumbo Red, Stump of the World, Cherokee Chocolate, and Cherokee Purple

At least one hour before serving, combine all the ingredients in a bowl. Before serving, be sure to taste for heat level and make adjustments as needed. If you've made it too hot, add more chopped tomatoes and scallions to dilute the fire.

White tomatoes are a genuine garden curiosity. Many people find them to be quite mild and sweet tasting. Others note that eating them slightly under-ripe, which ensures the presence of a tart element, intensifies the flavor. When at optimum ripeness, the color of the white varieties is more accurately described as ivory (white with a slight yellow tint), and the yellow coloring is more pronounced at the tomato shoulder. Fruit that grow inside the plant, hidden by foliage, could end up more white than those that are exposed to the sun.

Tomatoes that ripen to more than one color are fascinating to observe as they pass from formation to development and through to full ripeness. Tomatoes with stripes tend to be obviously different from the start, typically showing distinct vertically oriented, jagged stripes of dark and pale green. Over time, the true coloring starts to emerge, which is quite complex. All color combinations seem to be possible, and the colors change and deepen as the varieties move from partial to full ripeness. The swirled and marbled varieties don't show their true colors until fully ripe; often the red component of the red and yellow bicolored types deepens with time. The interior colors of the striped and marbled/swirled varieties could be uniform or a combination of several colors. My only advice for these types is to experiment picking at different stages and see what you like best.

The most complex, and often superbly delicious, varieties are those that retain their green coloring when fully ripe. My seedling customers often have to be prodded to try the green-when-ripe types, simply because they're confused about when to pick them. Two types of tomatoes retain green flesh at ripeness: those that develop yellow skin (which are easier to judge for ripeness), and those that keep clear skin (the true mysteries of the tomato garden).

GREEN WITH YELLOW SKIN. Those with yellow skin, such as Cherokee Green and Evergreen, scream out their ripeness by appearing as rich amber yellow. The surprise comes in the slicing, when the glowing green interior is exposed.

GREEN WITH CLEAR SKIN. As for the clear-skinned green varieties, such as Green Giant and Aunt Ruby's Green, I've learned that it is quite possible to determine ripeness by observing them, but it takes practice and experience. When at optimum ripeness, these types develop a very pale pink (quite lovely) blush at the blossom end. Of course, the tomato will yield a bit to a gentle squeeze. Once you grow these clear-skinned green varieties for a

THE EVOLUTION OF TOMATO FLAVOR

Tomato fruit flavor evolves and changes as the season elapses. Some people find the first-picked tomatoes are the best tasting; most find that it is the main crop that tastes the best. Most seem to agree: tomatoes that come on late, after the weather cools (especially at night), just don't have the flavor of the early or main-crop fruits.

Flavor in a particular variety can vary widely from season to season (certainly influenced by the weather any given year). It can also vary by specific location and specific cultural practices, such as fertilization and location (ground or container). Any disease that affects a plant will negatively influence flavor. And maybe just as important is the relationship between perception (what we see and our expectation) and what we taste. All avid tomato gardeners develop biases over time. We probably hold tight to our early favorites and become more resistant to admitting that recently found and tried varieties can possibly meet our expectations and match our experiences of those first few heirloom tomatoes grown long ago.

Swirls & Stripes
RED ZEBRA
LUCKY CROSS
GREEN ZEBRA
STRIPED GERMAN
RUBY GOLD

few seasons, you will find that you can tell when they are ripening by their appearance from a distance, if you are really observant. Though hard to describe, the ripe ones just look different — still green, but warmer than the cold-looking green of unripe tomatoes.

STORING RIPE TOMATOES

If you grow more than a handful of tomato plants, and you grow them well, the eager, frustrated anticipation that makes that first tomato take forever to harvest will soon be replaced with the panic of being overwhelmed with tomatoes. You'll be asking yourself a number of questions. How long can they sit on the kitchen counter? Should I put them in a bowl or line them up single file? Should they sit stem side up or down? Should they go into the refrigerator, the garage, or outdoors on the picnic table?

Large, ripe tomatoes, such as these fine specimens of Cherokee Purple, should be stored stem end down.

DON'T REFRIGERATE. The jury is in, and the vast majority of cooks, gardeners, and tomato aficionados agree that a tomato should never see the inside of a refrigerator. Two things seem to happen to a tomato once it is refrigerated. The texture changes, softening to what can best be described as "mushy." In addition, many of the chemical compounds responsible for the aroma and flavor of tomatoes are most likely highly volatile and are best experienced at higher temperatures. Chilling surely reduces the ability of many of the flavor elements to reach your nose and taste buds, leading to a flat, simpler taste experience.

STEM SIDE DOWN. An article in *Cook's Illustrated* magazine from 2013 concurred that it is far better to store tomatoes stem side down. The testing staff reasoned that the fruit stays fresher when any entry for air at the stem end is inhibited. They do, however, say that if a bit of stem remains attached, it is best to store the tomato stem side up.

THROUGH THE SEASON IN OUR GARDEN

Following are some of the lessons we've learned about dealing with the tomato harvest, especially when ripe tomatoes really start to come in at a rate far greater than anyone's ability to deal with on a daily basis.

PICKING UNDERRIPE TOMATOES

Full-size tomatoes can be picked when underripe and stored in such a way that they will ripen to the expected colors. The last fruit to set that reach full size and need to be gathered before frost can be stored in paper bags with an apple. Ethylene gas naturally emitted by the apple will act as a ripening agent for the tomatoes. Though they will reach full color, the flavor and texture of green-picked, indoor-ripened tomatoes are typically inferior to those ripened on the vine. However, fruit picked just shy of fully ripe (perhaps before a heavy rain, for example, or to avoid damage from critters) will ripen on the shelf within a few days and will possess most of the flavor of vine-ripened fruit.

Cherry Tomatoes Kick Off the Season

It usually happens that one day I walk into the garden and find that there's a single ripe cherry tomato. The first question arises: "Should I eat it, or bring it to my wife?" That first ripe cherry tomato seems to be a sort of cosmic trigger — the next day, or the day after that, two or three more are waiting. The production of ripe fruit accelerates, and soon it isn't a question of whether to eat or share; it gets to the point where I always bring a small bowl with me on my garden walk. Salads certainly become very pleasant now that a supply of fresh, ripe cherry tomatoes enhance the ingredient list.

They're Getting Larger . . .

Once the size of the ripe tomatoes increases, the cherry tomatoes may be forgotten. When the medium-size tomatoes begin to ripen, the kitchen counter lineup begins. Skewering, grilling, and slicing become possible, and the cookbooks, recipe cards, and Internet recipe searches become part of the daily ritual. The medium-size, round tomatoes typically yield heavily, and certainly take center stage from the cherry tomatoes, though we still enjoy snacking on bowls of cherries throughout the season.

The Heavy Hitters

Finally, the big ones arrive. The most enjoyable part of the tomato harvest season is when the sumptuous, colorful, large-fruited varieties begin to ripen, and we harvest big baskets of Brandywines, Mortgage Lifters, and Cherokee Purples. The kitchen counters are now cleared, so that the tomatoes can be clustered and lined up on every available space. We stand back and admire our bounty.

And then it hits us. We've gone from drooling, eager anticipation to dealing with the daily harvest to being utterly overwhelmed.

We've found that the large-fruited tomatoes simply don't hold up for as long once they're picked. They ripen very quickly, and we often

Most large-fruited heirloom tomatoes ripen gradually in a single cluster, and they also can exhibit great size variation, as seen in the cluster of Ferris Wheel shown here.

find that a half-ripe tomato picked one day, lined up next to its buddies, appears to be fully ripe the next day. We become aware of a foul odor and a puddle of liquid; sometimes tomatoes have internal issues (a worm going to work on the flesh, or internal blossom end rot), and these manifest themselves as pools of noxious rotten tomato liquid. Fruit flies are quickly drawn to such puddles as well, so we regularly monitor the condition of our fragile, ever-changing tomato collection. Wasting a single, perfect, lovely tomato seems like such a crime. That's when it's time to share!

Payoff Time

Despite being a bit overwhelmed by our harvest, we relish the first day of a given garden season when we can overlap slices of Brandywine, Cherokee Purple, Lillian's Yellow Heirloom, Green Giant, and Kellogg's Breakfast with fresh mozzarella cheese, drizzle with extra virgin olive oil, and top with some shredded

Ten Tastiest

BRANDYWINE

BRANDYWINE is a tomato that earns its lofty reputation. It is maddeningly inconsistent year to year, but when it does well, it is the most magnificently flavored tomato in my garden. Akin to other tomatoes on my list, it is a tomato lover's tomato, as the flavors explode in the mouth, sweetness and tartness exquisitely balanced.

basil leaves, a dusting of Parmigiano-Reggiano cheese, and a few grinds of black pepper and call it lunch or dinner. Great tomatoes taste best when you eat them in their most basic, unadorned form. What makes the experience even more precious is the knowledge that the delicious, large-fruited heirlooms can be the most finicky in terms of consistent yield and flavor, as they seem to be the variety type most affected by cultural or seasonal conditions. If it's a decent season, though, we quickly find that we are inundated with tomatoes.

COPING WITH TOMATO OVERABUNDANCE

The perfectly ripe tomato in the peak of its always-too-short season is the most poignant of fresh produce conundrums. Over the years, we've discovered some good ways of dealing with tomato overabundance — share with neighbors, hold tomato tastings with family and friends, and hunt through the recipe cards and cookbooks to find as many ways to enjoy them as possible. This is also the perfect time for seed saving (see page 133) as well as preserving your abundance for later (see page 126).

Cherry Tomatoes

Knowing that the cherry tomatoes will start the harvest and then quickly go into overdrive production, consider them the working-in-the-garden treat to keep the plants under control, and if you manage to have a bowl of them to bring into the house, use them in salads, skewer and grill them, or make the wonderful and unique Cherry Tomato Pesto described on page 120. Go through the bowl frequently and remove any that are cracked; in fact, during the picking process, toss all of the cracked ones, as they will just hasten the ripening (and overripening) of their bowl companions. Be sure to save seeds from those first-picked varieties on a given plant; because cherry tomatoes are quite seedy, just a handful of cherries will provide all the seeds you need.

Medium-Size Tomatoes

Pick medium-size tomatoes (4 to 8 ounces) in a range of ripeness, from half-ripe to fully ripe, so that you'll have tomatoes for immediate use as well as a few days down the line. This type of tomato — ranging in size from the golf ball of Tiger Tom and Green Zebra to the tennis ball of Eva Purple Ball — can be a bit awkward to use, being too small for a sandwich, but too big for use in a salad without cutting them into pieces.

I've found that as long as they are crack free, medium-size tomatoes last quite a long time post-harvest sitting on the kitchen counter. We do indeed cut them up for salads, but they are also wonderful brushed with a bit of olive oil and grilled whole, halved for use in tomato sauce, or canned.

Large-Fruited Tomatoes

I've learned from experience that it is a good idea to pick most of the really large tomatoes, often the most treasured varieties of all, at 50 to 75 percent ripe, rather than let them go fully colored on the vine. Since these often take the longest to ripen, can sometimes be the most vulnerable to lower yields (due to flowering in conditions not ideal for fruit set), and contain, as a group, many of the real flavor stars, it is important to get them safely harvested before a bird, deer, or some other hungry critter gets to them. Picking them a bit prior to full ripeness also provides more time to enjoy them, as fully ripened specimens go from perfect to nearly rotting quite quickly.

BEST USES FOR GREAT TOMATOES, LARGE AND SMALL

Because of their small size, cherry tomatoes typically ripen earlier in the season and provide the first chance to use homegrown tomatoes in favorite recipes. Their small size also makes them flexible to use, and because they're so prolific, you know you'll have plenty. Later in the season, the real culinary stars of the summer ripen — the large-fruited tomatoes. It is with these treasures that the flavor nuances emerge. And the rainbow of colors provides seemingly endless possibilities in the kitchen.

Cherry Tomato Pesto

Our standard recipe for traditional pesto included lots of fresh basil combined with olive oil, garlic, pine nuts, and Parmigiano-Reggiano cheese, all pulsed in a food processor to make a thick paste. It never occurred to us to use tomatoes in the preparation of pesto until seeing a recipe showcased in an episode of *Lidia's Italy* cooking show. By substituting cherry tomatoes for some of the basil, swapping toasted almonds for the pine nuts, and omitting the cheese, something entirely different emerges. We use the cherry tomato pesto just like the standard type: ample quantities tossed with cooked pasta, festooned with Parmigiano-Reggiano.

Any intensely flavored cherry tomato will work wonderfully in this recipe. With Sun Gold tomatoes, the result is a lovely pale orange. Black Cherries, Green Grapes, Snow Whites, Sweet Millions, Galinas — one could picture a plate laid out with splashes of different cherry tomato pestos, like a painter's palette, for a gorgeous and artistic meal.

Couscous Salad

This is a lovely-looking dish that is perfect for a rainbow selection of cherry tomatoes (or any tomato cut into bite-size pieces). The success of this dish relies on the intensity of the flavor of the tomatoes, so be sure to use your best. This dish tastes wonderful prepared several hours in advance and then allowed to chill in the refrigerator for a few hours. We've served it as a main dish for lunch or a light dinner, with or without bread, or by itself or as a salad over lettuce. The flavors are clean and bright, though savory from the garlic and paprika, with as much fire as you desire, which you control with the amount and type of hot pepper.

The dish involves tossing freshly prepared couscous (small grained, not the larger Israeli couscous) with cubes of roasted sweet peppers, diced cucumber, minced jalapeño or serrano peppers, and lots of halved sweet cherry tomatoes. Consider using a mix of colors, flavors, and sizes, including yellow varieties such as Lemon Drop, Galina, or Egg Yolk, the rich purple Black Cherry, pinks such as Sweet Quartz or Dr. Carolyn's Pink, the ubiquitous Sun Gold, and maybe some red Tommy Toes or Sweet Millions. Toss all the ingredients together and then mix in a dressing of olive oil, minced garlic, salt, pepper, lemon juice, ground cumin, and sweet paprika.

Caprese Salad

This dish simply layers thick slabs of fresh tomatoes with fresh mozzarella cheese, topped with basil, a drizzle of high-quality extra virgin olive oil, a few coarse grinds of rock salt and black pepper, and a dusting of grated Parmigiano-Reggiano. It reminds me of a great classical violin or piano concerto. The tomato is the solo instrument and takes (and deserves) center stage, but it is enhanced by all of the other ingredients/instruments. And, as with music, it is not too hard to tell if something is out of tune. To me, this is the ultimate test for proclaiming the relative quality level of a particular tomato variety.

This is a dish we reserve for the harvest stars of the tomato garden, including Brandywine, Green Giant, Cherokee Purple, Lillian's Yellow Heirloom, Great White, German Johnson, Dester, Cherokee Chocolate, Cherokee Green, Yellow Brandywine, Nelson's Golden Giant, Nebraska Wedding, Andrew Rahart's Jumbo Red, Nepal, Aker's West Virginia, Lucky Cross, Black from Tula, Stump of the World, and Polish.

TOMATO PALOOZA

A bountiful harvest of tomatoes is a cause for celebration and a catalyst for impromptu, or even more organized, get-togethers. This is especially true if lots of varieties are available to sample, and they are unfamiliar, interesting, or unusual. Nothing draws a crowd like a tomato tasting. The real value in having others around while sampling the harvest is in obtaining a range of opinions, since each person possesses unique sensory abilities, as well as preferences.

During our annual seedling sales, we're often asked about the differing flavors of our many selections. In 2002, a longtime seedling customer, and now gardening friend, Lee Newman, and I decided it would be fun to gather a small group of my customers for a midsummer get-together at a local park, where we would pool our harvests and have an informal tasting. Lee came up with the name Tomatopalooza, and it quickly became an eagerly anticipated and popular annual event. Starting off with just a few dozen home-gardening tomato enthusiasts, the event swelled to more than 100 people, providing an opportunity to taste, at no charge, up to 200 different tomatoes.

Rather than giving each grower their own station, we arrange the tomatoes by color on long tables. Our volunteer staffers cut the tomatoes into pieces during the event, to keep them from developing off-flavors from waiting in the summer heat. We taste, take notes, and hold discussions.

Over the years of putting on Tomatopalooza, we've discovered a few things:

- Though not always **logistically possible, starting with the whole tomato and cutting pieces just prior to eating provides the fullest learning experience, since the actual tomato color becomes difficult to judge using pieces alone.**
- Tomatoes **lose quality quickly after cutting, especially on a hot day, and tomato pieces that were pre-cut hours before often have off-flavors, thus making it impossible to fairly judge the various types.**
- We tend to eat **with our eyes, and our impressions of tomato flavor are often swayed by the color of the fruit.**
- Larger pieces **provide a much more accurate representation of flavor than very small pieces, and all components of the fruit — skin, seeds, gel, and flesh — should be present in each tasting bite.**
- Tomatoes **tasted first in any large tasting often are judged to be among the best of the tasting, probably due to palate fatigue toward the end. Keeping any individual session limited to 20 varieties or fewer will provide a truer analysis of each variety, especially if an opportunity to return to and retaste varieties exists.**
- Providing the ability **for tasters to write notes during the session is very helpful, as impressions easily get blurred and lost in larger tastings.**
- The flavor **of a particular variety can vary widely from season to season, throughout a season, and because of how and where it was grown, just to name a few of the many variables. If one fruit of one variety grown once doesn't impress, and it has a good reputation, it is important not to discount it as inferior for keeps.**
- People's abilities **to sense sweet, bitter, sour, and tart vary widely, and a single tomato at a tasting may come across very differently to each taster.**

6

Saving FOR THE FUTURE

WE GARDENERS OF TODAY OWE a great debt of gratitude to Mother Nature for coming up with such a wonderful edible whose seeds are so simple to save, and most often in an uncrossed state. Of course, our current bounteous selection would not exist if it weren't for those who came before us with the foresight to save seeds of so many wonderful varieties, as well as the seed-saving organizations and seed companies that work alongside gardeners, providing archives for the varieties and opportunities to share and purchase them.

We all have the choice to preserve our bounty — both by putting up tomatoes for the colder months to come and by saving seeds of special varieties for future gardening seasons. Neither process is difficult, and both are rewarding in countless ways.

PRESERVING THE HARVEST

Despite all of our best efforts, a successful tomato garden will produce more delights than can be eaten or given away before going bad. Fortunately, the flavor of summer can be preserved for the coming chilly days when one can only dream of the garden. A gardener who is armed with a brief list of preservation methods will be handsomely rewarded with quantities of frozen tomatoes for soups and stews, chewy and intensely flavored dehydrated tomatoes for pizzas and omelets, and multihued quart jars of delicious juice and rich, thick sauce.

Whole Frozen Tomatoes

Let's start with the easiest way to deal with an abundance of delicious tomatoes. The entire procedure is as follows: Check fresh-picked tomatoes of various shapes, sizes, or colors to ensure that they have no bad spots or spoilage. Wash and dry them, put them in freezer bags, zip to seal, and place them in a single layer in the freezer. That's all there is to it.

Frozen tomatoes lose the firm texture of garden-ripe fruit but retain all of the wonderful flavor. The best use for them is in cooked preparations, such as soups, stews, and sauces. The skin will easily slip off if you run the frozen specimens under water for a brief period. After you remove the skin, it's easy to remove the core (tomatoes don't freeze solid like ice), after which you can quarter them and add them to your favorite recipe. The only drawback is that you'll need ample freezer space, especially if you intend to grow and preserve a lot of tomatoes.

Milling

A quick and easy way to process large quantities of tomatoes is with a good-quality tomato strainer, such as Victorio or Squeezo. To convert the tomatoes into strained thick tomato juice, place raw tomatoes, fresh picked from the garden, into the collector at the top, and turn the crank (or, if motorized, turn it on). The seeds and skin are extracted from the juice and easily collected.

We used a Victorio strainer to process tomatoes for use in sauces during many of our initial gardening seasons, primarily when we grew red tomatoes such as Roma and Better Boy. Once we moved on to the colorful heirlooms, we transitioned to preparations where the individual colors could be retained separately. Even when we don't mind blending colors, our preference has evolved for chunkier-type sauces, so we don't often use a strainer these days. Still, if tomato juice or smooth sauces are your targets, nothing is quicker to use than tomato-milling equipment.

Processing tomatoes this way confirms the fact that they are as much as 95 percent water. Meaty Roma-type tomatoes, which have a much higher ratio of flesh to seeds and gel, produce the thickest juice. Many of the recipes for homemade tomato juice call for cooking the tomatoes down before putting them through the strainer. The same goes for tomato sauce: if you process the tomatoes raw, the juice will need to be cooked down over low heat or in a low oven to reach the desired level of thickness. The alternative is to cook down the raw tomatoes first, then put the cooked tomatoes through the mill. The juice or sauce produced can be frozen or canned (following the recommended canning guidelines on pages 130–131) for later use.

Dehydrating

Dehydration is a wonderful tomato preservation method. We adore the flavor of sun-dried tomatoes, particularly as a topping for homemade pizza or as an addition to omelets and frittatas. Anyone who has purchased sun-dried tomatoes, whether dry or packed in oil, realizes that the bits of super-intense tomato flavors come at a relatively high price. Thankfully, producing dried tomatoes at home is quite simple, either on baking sheets in a low-temperature oven, or with a dehydrator. The key to successful oven dehydration is using very low heat, allowing space

Ten Tastiest

CHEROKEE PURPLE

CHEROKEE PURPLE defines the ideal intersection of sweetness, tartness, depth, and texture. It is a tomato lover's tomato, finding a home in salads, on sandwiches, or just sliced into thick slabs. Cherokee Chocolate and Cherokee Green share all the wonderful flavor aspects of this tomato, with the added benefit of having distinctly different colors.

between the tomato pieces, which are arranged cut side up, and careful monitoring. Patience is also required. Depending upon the moisture level, the thickness of the pieces, and the oven used, the duration of the process can range from 6 to 12 hours.

The type of tomato that you use will have a significant impact on the product. My first attempt at dehydrating Sun Gold cherry tomatoes in the oven produced intense and delicious flavor, but very small fragments, since cherry tomatoes often contain the least flesh of all and have very thin walls. Tomatoes that dry the best, offering an adequate return for the effort, are smaller plum or paste tomatoes, or slices of meaty larger-fruited types. The meatiest tomatoes (those with the least amount of water) make the most substantial dehydrated product.

The tomato pieces may be seasoned or processed plain. Once dried, they can be stored as is, in oil, or frozen, but it is important to follow any specific guidelines for storage on whichever recipe that you follow, to ensure you don't end up with spoiled product.

If you have a dehydrator, the process for making "sun-dried" tomatoes is even easier, because you don't need to tie up your oven for long periods of time. By simply following one of the recipes available in the booklet that comes with your dehydrator, you will have a nice yield of delicious tomatoes after eight or more hours of processing time. Be sure to monitor the tomatoes closely toward the end, to achieve the desired consistency.

Canned Tomatoes

The idea of canning vegetables hits different people different ways. If you grew up on a farm, and garden and harvest chores were a major part of your early life, the thought of canning may be one you choose to not revisit. If you are a busy person who can barely find time to make an occasional home-cooked meal, the time-consuming nature of canning could seem like just a bit too much bother. But if you're trying to become more self-sufficient, if you love to cook, or if you become depressed at the end of the gardening season by the loss of all the wonderful flavors, canning may be just the thing to bring a bit of that summer experience into the cold, dark months of late fall and winter.

When we first started canning tomatoes, we bought the bible of canning — *The Ball Blue Book*. The book says that it takes about 25 pounds of tomatoes to prepare seven quart jars of processed tomatoes. We discovered that 25 pounds doesn't seem like all that much when you look at the parade of tomatoes lined up on the counter in the peak of the season, but once the tomatoes are prepped into a large bowl, awaiting transfer into the jars, the quantity seems overwhelming. Then, once you actually open up a finished quart jar of tomatoes, there's some disappointment at how small an amount it seems to be. This is when you question the wisdom of undertaking the canning ritual at all.

Make no mistake — the effort that goes into making that seven-quart batch is not insignificant. Of course, the whole canning process can be quite fun, especially if you can share it with someone and divide out the particular tasks. (My wife and I have become quite proficient at successfully canning without any major collisions or strife!) And, for me, there is nothing like grabbing a colorful quart jar of mixed heirloom tomatoes for a favorite recipe in the dead of winter.

ROASTED TOMATO SAUCE

Roasting your own tomatoes creates a sauce with amazing flavor and quality — especially compared with any sauce you'll find at the grocery store — and the ease of preparation is just plain silly. All it takes is a couple of large roasting pans and any seasonings you'd like to add (we like to roast ours with peppers, onion, and garlic). Once the sauce is done, you can either freeze or can it. The culinary beauty of roasted tomato sauce is in its intensity and complexity; roasting brings out deeper, richer flavor elements you won't find in sauce made on a stovetop.

INGREDIENTS

- 2 tablespoons olive oil
- 2 sweet peppers, seeded, cored, and cut into 1-inch cubes
- 1 large sweet onion, peeled and cut into ½-inch cubes
- 2 cloves garlic, peeled and minced
- 20 pounds (at minimum) tomatoes, all sizes and colors, cored and cut into 1- to 2-inch pieces
- ½ teaspoon red pepper flakes
- Freshly ground black pepper
- 1 teaspoon salt

1. Preheat the oven to 300°F.

2. Divide the oil, pepper, onion, and garlic between two roasting pans.

3. Add tomatoes until within an inch of the top of each pan; season with pepper flakes, black pepper, and salt, and stir.

4. Place the pans in the oven and cook for at least two hours, stirring occasionally. The sauce will bubble, reduce, and char a bit in places.

5. Either add more tomatoes to each pan and continue to roast, or, if the sauce is of a thickness that you like, remove the pans from the oven. If you wish to add some fresh herbs, this is when to do so; we've used torn leaves of fresh basil and/or oregano.

6. For a smoother sauce, transfer to a bowl and blend with an immersion blender, or leave as is for a coarse-textured sauce. Use the sauce immediately, or cool for further processing.

7. Can, if desired, according to instructions on page 130. Canned roasted sauce will last for up to a year, though it is best in the first six months after canning. We also fill freezer bags and freeze the sauce; it stores well for up to a year.

FAVORITE USES FOR ROASTED TOMATO SAUCE

Though we love using our roasted tomato sauce the day we make it, the true value is in incorporating it into recipes long after the garden is done. Of course, we love it ladled over cooked pasta, but we enjoy it even more when combined with eggplant. Because of the prolific nature of eggplant when grown in containers, we are often overwhelmed and needed an easy way to preserve it for winter use. By simply peeling eggplant, slicing into slabs, dipping into beaten egg and milk, dredging in bread crumbs, and baking on cookie sheets at 400°F until browned, the excess eggplant harvest can be squirreled away in freezer bags and popped out at a moment's notice for a great off-season meal. We use the baked, breaded eggplant slices in two ways with our roasted sauce. They provide the perfect layers for a traditional eggplant parmesan (alternating tiers of the eggplant with mozzarella cheese and the sauce). Even easier (and lighter) is simply placing the frozen rounds on a cookie sheet and crisping in a 400°F oven; they then serve as a bed for cooked spaghetti, over which the hot roasted tomato sauce is ladled.

PROJECT

HOW TO CAN TOMATOES THE LeHOULLIER WAY

WE ARE AVID CANNERS. Though we started with the *Ball Blue Book* and still follow its major principles and steps, we've made a few changes. Since we don't mind seeds in our canned tomatoes, we don't go to extremes to remove them all. I also combine seed saving with tomato canning, so most of the seeds end up fermenting in cups instead of in the jars of tomatoes, anyway. We also don't mind tomato peels in our cooking, so we don't go to the trouble of removing them when we're canning.

Whichever recipe you choose to follow, there are a couple of critical steps. You must thoroughly wash the canning jars, lids, and bands to avoid contamination. It's also important to warm up the jars before you fill them with hot tomatoes, so that the jars don't crack. We fill the canning jars with water just off the boil while we do our prep work.

1. Gather approximately 25 to 30 pounds of ripe tomatoes. Use a mix of whatever is ripe at the time, but avoid cherry tomatoes, since they are mostly seeds and juice. A variety of textures, colors, and flavors results in an attractive, delicious final product. Check the tomatoes carefully for areas that indicate spoilage and discard any that are unsuitable.

2. Cut the tomatoes in half and remove the seeds, if you're saving them (see page 133). Cut them further into chunks no larger than 3 inches wide, so that they'll fit into the mouth of the canning jar. Remove the tomato core and any cracked or blemished skin (toss them into a bowl for collecting compost). Place the good tomato pieces in a large, clean bowl.

3. Fill the canning pot halfway with water. The full quart jars of tomatoes will raise the level significantly; the level can be adjusted later by adding or removing water, as needed. Start the water boiling in the canning pot and have additional water boiling in a kettle.

4. Wash all of the jars, bands, and lids in hot soapy water. Fill the jars with near-boiling water, and place the bands and lids in a clean bowl filled with near-boiling water.

5. One by one, empty the hot water out of each jar. Add 1 tablespoon of lemon juice and 1 teaspoon of salt per quart jar, and then scoop the cut tomatoes into the jars with a clean ladle. Keep handy a well-washed wooden utensil as an aid to push the tomatoes into the jars, swishing it around the sides and throughout to eliminate air spaces. It is fine to crush or damage the tomatoes a bit; quite a bit of juice will be released.

6. Once each jar is filled to within half an inch of the top, use a clean cloth to wipe away any seeds or pulp or anything else that will prevent a seal between the lip of the glass jar and the rubber of the lid.

7. As each jar is cleaned, place a lid on top of each and screw the band on securely, but not too tightly.

8. Carefully place the quart jars into the wire canning rack with a jar-lift tool. Then carefully lower the basket into the boiling water of the canning pot. The quart jars must be completely immersed with at least an inch of water over the jar tops. More boiling water can be added (or removed) to adjust the level as needed. Place the top onto the canning pot, and once it returns to a boil, lower the heat to simmer.

9. Forty-five minutes later, carefully remove the jars and place them on a cooling rack. Within 5 minutes, you will begin to hear a popping sound as the vacuum is formed in each jar, confirmed by a slight depression in the center of the lid. Any quarts that do not seal because of detritus between the seal and the jar can be opened, the offending solid matter removed, and the jars reprocessed for an additional 30 minutes. If you choose not to do this, refrigerate the tomatoes that didn't seal the first time and use them within a week.

10. Once the jars are cool, remove the bands, wash the jars and bands well, replace the bands if desired, and store in a cool, dark place such as your pantry or food closet. Be sure to write the processing date on each jar.

Once the quarts are processed, you will find that the tomato solids rise to the top and a layer of clear liquid forms. This is a completely normal testament to the fact that tomatoes are mostly water. The more solid tomatoes you use for the canning process (such as Roma types or the heart-shaped varieties), the less liquid will form. We find that canned tomatoes keep their quality and are safe to eat for up to a year or more. One thing is certain: once you realize how delicious and versatile these tomatoes are, you will never be able to can enough.

Ten Tastiest

MEXICO MIDGET

The first time I don't plant **MEXICO MIDGET** in a large container in our driveway garden may be when my wife, Susan, decides to put me in the doghouse. If it were possible to squeeze the rich, complex flavor of the large beefsteak-type tomatoes into a package the size of a pea, Mexico Midget would be the result. Though it is not all that well known or widely grown, it is one of the most sought-after varieties with my seedling customers and friends. This truly addictive, luscious variety proves that great things can indeed come in tiny packages.

SAVING SEEDS

The best time to save seeds from your open-pollinated tomato varieties is when the first tomatoes ripen. I use the first ripe tomatoes for seed saving to reduce the chances that the bees will have worked the flowers and cross-pollinated the seed in the tomato. (See page 160 for more on isolating flowers for the purposes of seed saving.)

Tomatoes that are in edible ripe condition are best, allowing for the pairing of seed saving with eating, cooking, or preserving. Since the genetic material in every seed on every tomato of a particular plant (as long as it is a stable, non-hybrid variety) is the same, it isn't necessary to use the most perfect specimens on a given plant. (The single exception to this is if the bees that visited the flower cross-pollinated it with another variety; seeds in that specific tomato would produce a hybrid of the two varieties.) The most important thing to watch for is that the tomatoes on a given plant are true to type — that they match your expectations from either the description or past experience.

Harvest for Both Seed Saving and Eating

When you're harvesting ripe tomatoes for eating, be sure to think about seed saving, too. I carry a permanent marker with me whenever I harvest from the garden. After wiping any moisture off the harvested tomato that I intend to save seed from, I jot down the name or reference number directly on the tomato. Later, after the picking is done, my family knows that any tomato with writing on the shoulder near the stem attachment spot is reserved for seed saving. Cherry tomatoes are too small to write on, so I collect them in a plastic cup and write the name of the tomato directly on the cup.

Seed saving is so simple, yet there are also many differing opinions on best practice. Here are the three general ways to save tomato seed:

- Drying unfermented seeds on an absorbent material, such as newspaper or paper towels
- Fermentation, followed by drying on paper plates or another absorbent surface
- Chemical treatment with no fermentation, followed by drying

Drying Unfermented Seeds on Absorbent Material

This is a very simple way to save small quantities of tomato seeds. Cut the ripe specimen tomato in half and gather some of the seeds. You may wish to put them in a sieve and rinse them a bit, pushing the mass against the side of the sieve to clean off as much of the pulp as you can. Either way, spread the seeds thinly, in a single layer, on an uncoated paper plate, a piece of newspaper, or paper towels. Over a week or so, the seeds will dry and adhere to the drying surface. Peel off the seeds and store them as described beginning on page 136.

There's no doubt that this method is widely used because of its simplicity. In my early years of seed sharing among SSE members, I often received samples of seeds folded into homemade paper packets, which contained small samples of seeds stuck to bits of newspaper, napkins, tissues, or paper towel. One issue with this method is that any diseases on the seed surfaces that are typically removed by fermentation would carry through to the dried seeds. If you're saving large quantities of seed, there is also the cosmetic issue of paper bits stuck to the seeds (though this won't cause problems with germination).

Fermenting

I find fermenting tomato seeds easy and effective, though a bit odiferous. The advantage is that it helps remove any pathogens on the seed surface, and below or in the surrounding seed gel. The seeds end up clean, attractive, and easy to package and store.

Fermenting also removes the natural germination inhibitor that coats tomato seeds, leaving the seeds vulnerable to germination. That's why it's important to limit the amount of time the seeds are fermenting; otherwise, you would end up with sprouted tomato seeds.

PROJECT

HOW TO SAVE SEEDS USING THE FERMENTATION METHOD

ARMED WITH YOUR PERFECTLY RIPE NON-HYBRID/HEIRLOOM TOMATOES, you are ready to start your seed saving. Beware, however, that during the seed-saving process, mix-ups or errors could be introduced in many places. Consider this another plea for good documentation and careful discipline so that future tomato mysteries can be minimized.

Typically, when I have a quantity of tomatoes that are ripe and ready for not only seed saving but processing, I arrange them by variety and work on my kitchen counter with a cutting board, my knife, a big bowl, cups, a writing utensil, and my compost bin handy. Here is my seed-saving procedure:

MATERIALS

- **Medium-size cups that can be written on**
- **Permanent marker**
- **Sharp knife**
- **Bowl or pan, if the edible portions that remain will be used for a recipe**
- **Paper towels**
- **Fine-mesh sieve**
- **Spoon**
- **Uncoated paper plates**
- **Storage containers for dried seed**

1. Note the variety of the tomato and write it on a cup, along with the date. Cut the tomato in half and squeeze each half over the labeled cup to capture the seeds and gel. After squeezing, cut away the core and stem part and toss into the compost bin, and put the remaining edible flesh into a bowl for processing.

2. If the pulp in the cup is very thick or dry or can't be swirled, add just a bit of water. Check the knife and cutting board to make sure no seeds are present, and then move on to the next variety. Once seed is saved from all of the varieties, move the cups to a place where they can sit undisturbed for a few days, out of the sun. The smell will be pretty bad, so outdoors in a protected place or a garage works well. Cover the cups loosely with a paper towel. The paper towel should help keep flies that are attracted to the fermentation from laying their eggs in the pulp.

3. If the conditions are very warm (85°F or higher during the day), check the cups of fermenting pulp on the second day. If the weather is cooler, it could take up to 5 days for the fermentation to take place and the fungal layer to form. Once the pulp starts to ferment (indicated by an offensive smell and the presence of a layer of white, or occasionally black, fungus), it's time to rinse the seeds.

4. Remove the paper towel and bring the cups of fermented tomato pulp to a sink. Add water to each of the cups to within 1 inch of the top, then stir each cup a few times (be careful that seed from one cup is not introduced into the next). Allow the contents to settle for a few minutes. The good seeds will sink to the bottom of the cup, and fungal material and other solid debris will float.

5. Gently tip the cup over the drain and pour off the floating debris. The cup contents can be quite concentrated, viscous, and opaque at this point so be very careful to not allow any seeds to escape along with the upper debris. Add more water, swirl again, and pour off more floating non-seed material. Repeat until you have a cup of clear water and settled seeds.

6. Pour the material through a fine sieve (make sure the mesh is sufficiently small to capture the seeds). With the back of a spoon, push the seeds against the sieve under a stream of water to ensure any remaining non-seed solids rinse through. Take a cloth and push against the seed mass from the outside of the sieve to wick away excess water.

7. Label a paper plate with the variety, scrape up the seeds into a mass with the spoon, and press them onto the plate in a single layer. Store the plates of clean, spread, labeled seeds in one layer in a place where they can air-dry. After a week, the seeds should be sufficiently dry for final storage.

Seeds can be stored in a variety of ways. Glass or plastic vials and manila coin envelopes are my methods of choice. If your storage area is humid, airtight containers such as vials are best.

Saving Seeds without Fermentation

Some swear by this relatively new method of seed saving. The main ingredients are trisodium phosphate (TSP) and bleach. Follow the fermentation method on pages 134–135 up to step 2; you will have the labeled cups of tomato pulp and seeds. Add a 10 percent TSP/water solution to the seeds and pulp until the cup is approximately three-quarters full. Let the seeds soak in the TSP solution for 15 minutes, which will dissolve the gel and remove any pathogens. Pour off the top material and sieve and rinse the seeds, then return them to the cup. Add a 10 percent bleach solution to the seeds until the cup is about one-quarter full, and soak them for one to two minutes. Sieve once more, and use hot (120°F) water for the rinse. You are then ready to scrape the seeds onto the plates for final drying.

STORING SEED

The main enemy of tomato seed is moisture. Any seed storage option should take this into consideration. The other interesting thing I've learned about tomato seed is its surprising longevity, which is not something you would know if you read many commercial seed catalogs, whose success is based on bringing people back to purchase seeds each year. Over the years, I've stored my tomato seed in screw-top glass or snap-top plastic vials, where it has experienced all sorts of temperature changes as we moved about. Even given that rough treatment, I've found that my saved tomato seed germinated very well even at 12 years old, and occasionally up to 14 years.

Tomato seeds will remain viable even longer if stored in the freezer, in a vial with a small packet of silica gel. The main caution for using

seeds stored in the freezer is to allow the seed containers to reach room temperature before opening them. Otherwise, unwanted moisture from the air will be drawn into the containers and negate the work that was done to enhance the seed life.

I'm a very frugal gardener, and fortunately all of the stages of gardening can be done on any budget. My preference is to keep things as simple and inexpensive as possible, and that extends to seed storage. I've moved from the glass vials (which are not inexpensive) to plastic vials (which are less expensive but still a significant cost, especially when saving a few hundred varieties each year) to small manila coin envelopes, which I don't seal but just store upright in my office. Since we have a heat pump, temperature and humidity move in quite a narrow range, so this method works just fine.

One critical final part of the seed-saving process is documentation; a good system is very helpful in terms of locating the seeds you save, and ensuring that you know exactly what they are. I take it even further and aim for knowing each saved seed's genealogy, which helps me determine possible reasons for any unexpected surprises.

MEET THE VARIETIES

Tomatoes That Changed My Gardening Life

There is a special set of tomatoes that marked my entry into the wonderful world of heirloom varieties, and each played a role in converting me from exclusively hybrid to mostly non-hybrid types. Though they are not necessarily the best flavored, most colorful, or most unique, my tomato interest could not have evolved to where it is today without obtaining these varieties early on. Each is a tomato that was special enough to be maintained and passed on, and each possesses its own distinctive story.

Anna Russian

In 1989, the postman delivered a letter to me from Brenda Hillenius of Corvallis, Oregon. Inside was a letter written on delicate onionskin paper along with a small packet of seeds. The variety, Anna Russian, was given to Brenda's grandfather, Kenneth Wilcox, some years before by a Russian immigrant whose family sent him the seeds.

I grew Anna Russian for the first time in 1989 and was immediately enthralled. The plant itself was rather skinny, unimpressive, and not particularly vigorous, and it required frequent support to the proximal stake to remain upright. (All heart-shaped varieties tend to show this scrawny growth habit, but Anna Russian took the cake.) What was impressive, however, was the beauty and flavor of the tomatoes. Starting to bear quite early, Anna Russian produced nearly 20 pounds of fruit at an average weight of about half a pound each. Unlike many heart-shaped varieties, which can be quite dry, solid, and bland, Anna Russian was full of juice, with a succulent texture and full-bodied, absolutely delicious flavor that had all elements — sweetness, tartness, intensity, complexity — in perfect balance. This pink heart-shaped tomato is one that I return to often for reliable yields of superb fruit.

I was sure to save plenty of seeds that first year I grew it, and I listed it in the SSE yearbook. Many SSE members requested seed, all seemed to love it, and it found its way into seed catalogs, thus this wonderful tomato with a charming story is now widely available for gardeners.

Tiger Tom

One of the joys of becoming an SSE member was the opportunity to interact with real tomato enthusiasts via handwritten letters delivered in the mail, a format that is rapidly becoming obsolete. Among my first tomato variety requests as an SSE member was a set of tomatoes from Jim Halladay of Pennsylvania. Mr. Halladay sent me Tiger Tom in 1987, and it quickly became a flavor favorite. Jim noted that he obtained Tiger Tom from Ben Quisenberry (of Brandywine fame), and that Ben got it from a Czech tomato breeder. With its lovely striped appearance and snappy, tart flavor, Tiger Tom is a premier salad tomato that I continue to grow regularly.

Ruby Gold

The excitement I experienced when growing tomatoes of different colors for the first time is best represented with the variety Ruby Gold. It is one thing to read a description of huge tomatoes that end up as blends of yellow and red; it is quite another thing to watch it happen. I obtained it in 1987 from SSE member John Hartman of Indiana. Ruby Gold was first offered to American gardeners in 1921 by the John Lewis Childs seed company of New York. It appeared to go out of favor and was rediscovered at a West Virginia market in 1967 by Ben Quisenberry. Through the years, the same tomato seems to also have owned the names of Early Sunrise and Gold Medal.

Bisignano #2

When I received my first SSE yearbook in 1986 and the world of heirloom tomatoes presented itself, it was tempting to sample as many unusual colors as possible. As I read through descriptions, Bisignano #2 stood out as a red tomato worth trying. Supposedly, the tomato originated from Italy, and it was featured in the garden of a Mr. Bisignano, a finalist in the 1984 Victory Garden competition.

Unusual in that the fruit shape varies widely on any given plant (from oblate to nearly heart shaped), Bisignano #2 possesses an intense, truly deep tomato flavor that is at home in a salad, on a sandwich, or as a main component of sauce. It has been a favorite tomato for nearly 30 years.

Polish

Leaping out of the soil and producing vigorous seedlings with enormous potato-leaf foliage, Polish impressed me in every way. My initial experience growing Polish provided a valuable lesson about tomato colors. Described by long-time SSE member Bill Ellis as "brick red," it was actually a pink variety, having clear skin over red flesh. The flavor was just breathtaking, and Polish became the first of the knockout heirloom varieties for me. Though through the years I've located tomatoes that are the equal of Polish in flavor (including what I assume to be the authentic version of Brandywine), it will always hold a special place in my tomato-themed heart for the way that it opened my eyes to a completely new world of gardening and eating.

Yellow White

Reading of a tomato that was actually a wedding gift was intriguing; that it was white made obtaining and growing it irresistible as I set off on my heirloom tomato exploration. The full tomato name appears to be Kentucky Heirloom Viva, named by H. Martin of Hopkinsville, Kentucky, who donated it to the Landscape Garden Center of the University of Kentucky. Viva Lindsey was a friend of the Martin family; Viva was given this tomato, already an heirloom variety, as a wedding gift in 1922 by her husband's great-aunt. The medium-to-large, nearly white tomatoes have a lovely pale pink blush on the blossom end. The mild, sweet flavor is just lovely, and it is tomatoes such as this, combining unique beauty with a captivating history, that drew me so deeply into heirloom tomatoes.

Hugh's

A favorite period of my SSE membership was the early years, between 1986 and 1992, because of the activity and visibility of many of the elderly seed savers whose contributions essentially built the organization. Men such as Thane Earl and Fax Stinnett will always be special in the minds of tomato enthusiasts from those days. Another of those special people is Archie Hook from Alexandria, Indiana. Mr. Hook apparently had a small greenhouse in his neighborhood and provided seedlings for many of his gardening neighbors. One of the first heirlooms I tried was obtained from Mr. Hook, a truly outstanding variety called simply "Hugh" (from his letter), more commonly known today as "Hugh's." Mr. Hook indicates that he had this variety from the 1940s. The first time I grew it, 40 pounds of tomatoes were produced on a single plant, most coming in at well over a pound. The color of Hugh's fruit is very pale yellow, similar to Lillian's Yellow Heirlooms. The flavor of Hugh's has varied a bit for me through the seasons, ranging from outstanding and intense to mild and sweet. It's one of those varieties whose flavor seems to vary more than most, depending on a season's particular weather conditions.

Giant Syrian

When I started dabbling in seed swaps through the National Gardening Association magazine (at the time called *Gardens for All*), I met some wonderful, giving people. Few shared as many heirloom treasures with me as Charlotte Mullens, of Summersville, West Virginia. I already referred to the Mortgage Lifter that she sent me, but my favorites are Gallo Plum and Giant Syrian. Charlotte received the seeds of Giant Syrian from Harold DeRhodes of Ohio, who claimed that his family had grown them for a long time. I was delighted with my initial exposure to this tomato, as it was immense, heart-shaped, delicious, and unusually red (pink is more typical for heart-shaped varieties). Giant Syrian produces amply on very rampant, vigorous vines that show the typical spindly characteristic of just about all heart-shaped types. I will always think fondly of the late 1980s through the early 1990s as a time when there was great sharing of varieties between like-minded gardeners through seed swaps and the SSE. In a way, many were discovering the joy of seed saving and heirlooms at that time, building the foundation for a hobby that continues to grow each year.

7

Breed Your Own TOMATOES

OVER THE LAST 35 YEARS, growing tomatoes of readily available varieties has been highly satisfying. But at some point, my innate curiosity and love of science and discovery kicked in; it was clearly time to take things to the next level. I started simply with growing out seedlings with unexpected leaf shape and eventually ended up launching a project to create new dwarf tomato varieties. Taking on projects like this is fun, not difficult, and very rewarding.

THREE WAYS TO CREATE NEW VARIETIES

When reading old seed catalogs from the late 1800s, it's clear that the thirst for improved tomato varieties — stimulated by the needs of the customer as well as the need for seed companies to achieve a competitive edge — is nothing new. In fact, you can use the same "single plant" selection method that Alexander Livingston pioneered (see page 223) to create your own new tomato variety.

The other two ways to create your own tomato variety are dehybridization (a bit easier, at least at the onset) and starting with a purposeful cross (more challenging but more certain). For dehybridization, you start with a hybrid, such as Sun Gold, and using saved seed, you grow out and select interesting things to take forward and stabilize. For starting with a purposeful cross, you create a hybrid of choice in hopes that future generations of saved seed and selections lead to something new, different, and unique.

The Single-Plant Selection Method

VARIATION 1: FROM A POPULATION OF A SINGLE VARIETY

This method may or may not provide a good starting point for your work, because success depends upon sufficient variability or instability of your starting seed stock. It is therefore best for those who think they've found a distinctly different plant and have the room to grow out lots of plants to see if they indeed did.

Here is the process for working with a single variety:

- Conceive of an objective for a particular variety — for example, if you love Brandywine but find that it ripens too late, or is too crack prone, or your aim is for consistently larger fruit or higher yields (all attributes that have a genetic basis). Pick one to focus on, especially if you think you've seen signs of it in past plantings of that variety.
- Start sufficient seedlings to provide a real chance to see variation in the planting of your particular variety of interest — for example, you want to grow out 25, or 50, or 100 plants of Brandywine from that seed stock.
- Observe carefully and take notes throughout the season, looking for the plant or plants that are distinctly different and show the tendency toward the trait you desire. This is where there can be disappointment; if the Brandywine stock is well stabilized, it may be completely uniform and show little to no variation.
- If you find a plant or plants that show the desired attribute, save seeds from fruit on just that plant (if you are saving seeds from several plants, be sure to keep seed from each separate). These seeds are your starting point for next year.
- The following season, grow out a large quantity of the seeds from the selected plant, as well as a few of the typical Brandywine, to use as comparison. If the new selection shows the desired trait (probably not in all of the plants, but in increasing numbers), save seed as in the previous year.
- Repeat this process for a few more seasons until you can reliably produce plants that exhibit your desired improved attribute. Be sure to give the variety a distinct name, so as to avoid confusion with the original variety. Be creative and have fun; you just succeeded in creating a new variety of tomato.

VARIATION 2: FROM AN UNUSUAL SEEDLING OR FRUIT

A much more likely way to develop a new tomato variety is to start with something that is clearly different right from the start (in the seedling stage), or from the appearance of the first

harvested fruit. I successfully used this method to create Lucky Cross and Little Lucky (starting with a single regular-leaf seedling appearing in a large planting of Brandywine, a potato-leaf variety) as well as Cherokee Chocolate and Cherokee Green (from off-type ripe fruit of Cherokee Purple and Cherokee Chocolate, respectively).

Since "regular" leaves (those with serrated or toothed edges) are dominant over potato-leaf foliage, the appearance of regular-leaf seedlings in a planting of a potato-leaf variety could be from a stray seed or mix-up, a mutation (very rare), or an accidental hybrid (the regular-leaf seedling would be an F_1 hybrid between the potato-leaf variety and a nearby regular-leaf variety). Sometimes it is possible to decide what you are dealing with once the unexpected seedling is grown out and the fruits form and ripen, especially if the outcome is something quite unique. If this is not the case, then it is necessary to grow out seed saved from that unusual seedling to observe the expected variation and segregation that is indicative of a cross.

SEEDLING. Here is the process for working with a distinctive seedling:

- Carefully observe a planting of a particular variety and note any seedling that is distinctly different in leaf shape or color. An example is a few regular-leaf seedlings in a pot or cell of 20 potato-leaf seedlings (or vice versa).
- Grow one or more of the unusual seedlings to the fruiting stage, carefully observing if or how the fruit relates to what you'd expect from the variety you planted. If the fruit is different and desirable for any reason (size, flavor, or color), save seeds from the unusual tomato in preparation for growing the following year.
- Grow out sufficient numbers of the seeds from the selected tomato to allow for assessment of uniformity; in all cases, grow out as many as you have space for. If there is a mix of leaf types, grow some of each.
- When the fruit matures, carefully observe and document colors, sizes, and flavors. If no differences are present (the fruit from all plants are the same), you are likely working with stray seed of a known, named variety, and the selection project ends. If you have a diverse array of tomato types, develop objectives for the project. Will you carry forward one or many? Exactly what is the goal for each in terms of fruit size, shape, color and flavor, and leaf shape? The answers to these questions become the shape of the project going forward.
- At this point, this becomes a process of repeated selections each year until you reach stability with each of your objective tomatoes. It can take up to six generations or more to develop new stable varieties.

FRUIT. Here is the process for working with a distinctive fruit:

- If you find an unusual, unexpected tomato of interest (in my case, one of my Cherokee Purple plants produced brown tomatoes), be sure to save seeds from just the tomatoes on that single plant. Do a mental check, however, by thinking of other tomatoes you have grown, in case you grew a stray seed of a known variety. If that is likely the case, the project is done before it begins. It is wise to proceed only with something that is quite different from anything in your growing experience.
- The following year, grow out as many plants from the saved seed as you can manage.
- Once the fruit ripen, note diversity of size, color, and flavor. If the population is uniform, you likely identified a mutation, and your project is complete; because of serendipity, you are the proud parent of a new tomato variety. If the population is diverse, you are working with a cross, and it becomes a dehybridization process (see page 148).

MEET THE VARIETIES

Lucky Cross, Little Lucky

LUCKY CROSS

A chance cross can sometimes occur when a bee is traveling with pollen from plant to plant and catches open flowers at just the right time to interrupt the typical self-pollination act. In the case of the Lucky Cross project, my 1993 garden contained a Brandywine tomato planted next to an unusual striped experimental variety named Tad (a work in progress). I can only deduce that a bee must have visited a Tad blossom, picked up some pollen, and deposited it on a Brandywine flower, which then pollinated and set fruit. I saved lots of seed from Brandywine that year, and I mixed all of the seed from fruits from that particular plant into one storage container.

In 1997, I started about 20 seeds from the container that, unknown to me at the time, contained some seeds that were the crossing work of the busy bee. The vast majority of the seedlings emerged with the expected potato-leaf foliage for the first true leaves. However, a few seedlings had regular (serrated) foliage. Rather than rogue them out, I took a chance and decided to grow one in the garden to see what it produced. The three possibilities were that it was a cross, a mutation, or a stray seed or mix-up made in the seed-saving process.

The unexpected, regular-leaf plant formed large, oblate green fruit, and I was surprised to note the distinct light- and dark-green stripes on the developing tomatoes. This one observation confirmed that I was working out an unexpected cross. It couldn't be a seed-saving mix-up, because I didn't grow any oblate striped varieties. It was not likely a mutation either, because the striping was indicative of the Tad and Brandywine combination, and the real possibility that Tad pollen got into the flower from which the Brandywine fruit provided the seed I was growing. As the

LITTLE LUCKY

tomato ripened, I was delighted to find a large pink tomato with distinct vertical, jagged gold stripes. As a bonus, the tomato was delicious and possessed Brandywine's rich, full flavor. In fact, if I were to work out the possible appearance of a Tad and Brandywine cross, based on my very basic knowledge of tomato color dominance, this would be it.

What I failed to realize was the potential of the cross in terms of diversity of sizes, shapes, colors, and flavors. Finding and growing that single regular-leaf seedling led to work — growing, observing, and selecting — that continues to this day, with other promising new varieties showing themselves each time I dip into the line to grow a few.

After all, harvesting and saving the seed from the tomatoes from an unusual seedling is really the first step in this process, and it can take years from this point to develop a new tomato. However, it can be well worth the effort. I was fortunate to be able to share this project with a nearby garden enthusiast, Larry Bohs, and between us, we grew, discovered, and selected our way to two great new tomatoes. Within a few years, one of the favored new selections was essentially a yellow and red swirled, bicolored version of Brandywine itself. It has the potato-leaf foliage, large fruit size, and, most important, flavor of Brandywine, but with the unique coloring. I named that selection Lucky Cross. In parallel, a smaller-fruited version emerged, which we called Little Lucky. Thus was borne, out of a chance cross, the two best-flavored bicolored tomatoes that I know of to this day.

Dehybridization

Another way to create a tomato you can call your own is by starting with a known, commercially available F_1 hybrid. There are two main reasons to undertake this type of project: if you really love the hybrid and want to create an open-pollinated (non-hybrid, stable) version, or if you are just curious and want to see the richness (or lack thereof) of genetic possibilities in a hybrid. The former has a specific goal in mind, and the latter is completely open ended. Success is not guaranteed in either approach, but both can become a fun adventure.

Commercial hybrids are more complex now than they were initially. It is pretty certain that the first hybrid varieties sold as hybrid seed, such as Big Boy (released in 1949 by Burpee), were simple crosses of two non-hybrid varieties. With the advent of breeding for specific disease tolerance, hybrids started to become more complex in their parentage. What this means for the amateur gardener hoping to use dehybridization to find good new varieties is a better chance of finding a tomato more akin to the parent (which was likely selected more for its disease-fighting ability than its eating qualities).

Here is the process for dehybridization:

- Save seed from a few fruit of a hybrid tomato you would like to work with. These are F_2 seeds. Have an overall objective in mind: Are you looking to approximate the hybrid, or just to see how diverse and interesting the outcomes will be?

- The following year, plant some saved seed (F_2) and grow out as many seedlings as you want to work with — the more you grow, the more diverse fruit types you will have.

- Make careful observations and take notes. Did you find something particularly to your liking — either because it seems quite close to the hybrid (and your goal is to replicate it as closely as possible), or something really new and interesting emerged that is well worth following through on? Save seeds (which are now F_3-generation seeds) from tomatoes from specific plants that provide the leads to carry forward.

- Start the F_3 seeds and decide how many plants you wish to take forward. Grow the plants to maturity, evaluate the tomatoes that are harvested, and save the seed (which is now F_4 generation) from those that you wish to carry forward.

- Repeat seasonally until you have uniform, predictable results. The tomatoes are essentially new — your own creation — and can be named. Note that you often need to reach F_8- to F_{10}-generation seed to consider them stable, open-pollinated, new varieties.

Starting with a Purposeful Cross

All of the methods described until now rely to some degree on chance. This method is far more planned and reasoned, at least at the start, but then uses chance to finish the project. It's also the most skill-intensive, as it requires obtaining knowledge of carrying out successful tomato crosses. And although great results can be realized even in small gardens, the chances of finding something really great and name-worthy increase substantially with the number of plants that can be grown out each year.

Often, the goal of a cross isn't entirely clear, but perhaps the objective is just to see what happens when two favorite or interestingly colored tomatoes are crossed. The first finding will be the attributes of the resulting hybrid, which will show the dominant traits of each parent. The fun, of course, is seeing what occurs from saved seed, especially in the early generations when the diversity of the possibilities becomes evident.

Here is a summary of what is involved in creating a new variety on purpose:

- Select two parent varieties, taking into consideration what you are aiming for, such as color, flavor, ripening time, fruit size, or disease resistance.
- Grow several plants of each variety to ensure an adequate supply of blossoms and pollen.
- Decide which variety will be the pollen accepter (female) and which will be the donor (male). It can be helpful to use a variety with an obvious recessive trait for the female, as the success of the cross will then be easily seen when growing out the presumed F_1 seed, which will exhibit the dominant growth characteristics of the male donor.
- Select a few flowers that are at the appropriate age for removal of the anther cone and carefully remove the flower feature; the flower should be about to open, with medium yellow sepals, but still essentially closed.
- Collect pollen from the male donor by vibrating the flower over a collection vessel. Be aware that humid conditions can make pollen collection difficult.
- Using your finger or a brush or the collection vessel, dip the pistil of the "female" plants into the pollen. Repeat this daily for a few days. Mark the flower so that you know it is crossed, and record the nature of the cross.
- Watch for tomato development of the pollinated blossom. If it drops off, try again with a different flower.
- Save seed from the newly created hybrid.
- The following season, grow the new F_1 seed and note the characteristics of the plant and fruit. Save seeds from the tomatoes produced, which are F_2-generation seeds.
- Grow out as many of the F_2-generation seeds as you can fit into your garden, carefully observe the results, and save seeds from each plant, which are F_3-generation seeds. Clearly note which ones meet your project goals and focus on those going forward.
- Repeat this process for as many generations as it takes to reach stability. It could take as many as eight generations to produce a stable new variety. Name your new tomato.

All parts of the tomato flower can be seen clearly in this picture. This blossom exhibits an exerted stigma; the tip of the style extends beyond the anther cone.

THE DWARF TOMATO BREEDING PROJECT

The best and most current way to illustrate the purposeful cross method is to describe a unique project that I co-conceived and have been co-leading since 2006, called the Cross Hemisphere Dwarf Tomato Breeding Project. The idea for the project emerged during our annual seedling sales through conversations with our customers. Since the vast majority of colorful, great-tasting heirloom varieties of tomatoes are indeterminate in growth habit, the choices for those gardeners who can't find the proper location to grow them or have the physical ability to manage the rampant vines are severely limited. We strove to carry a supply of as many good shorter-growing varieties as could be found, but the selection of delicious determinate tomato varieties is uninspiring, and the little-known family of tomato varieties known as "dwarf" is even more limited in number.

Before our project, dwarf-growing varieties were available in just a few colors, sizes, and shapes. With my gardening friend Patrina Nuske Small in Australia handling most of the crosses, an all-volunteer, all-amateur, worldwide team was assembled, and we embarked on a really unique way to learn about tomato genetics together, the different growing seasons in our respective hemispheres allowing us to carry out two seasons of work in one calendar year. A tomato shown in the 1915 Isbell seed catalog, New Big Dwarf, demonstrated how we could succeed. By using the flavorful, colorful, large heirloom varieties as breeding partners with a few of the known dwarf types, we thought we could create some really special new tomatoes, given sufficient seasons of selection work following the crosses. The spectacular array of heirlooms available for Patrina to use as tomato breeding parents really opened up so many possibilities for our project, allowing inclusion of every available tomato color, shape, and size.

With the work beginning in 2006, the team succeeded in creating 17 new, unique, stable dwarf-growing varieties, the first of which was released through a selection of seed companies in 2010, with additional varieties following in 2011 and 2012. More than 150 people participated in the project leading up to those releases, and the project continues. Those seeking to grow dwarf tomatoes can now easily find large-fruited examples in red, pink, purple, brown, green, yellow, and white, with orange, swirled bicolors, and distinct stripes soon to follow.

Dwarf Blazing Beauty

Sean's Yellow Dwarf

Dwarf Sweet Sue

SAVING SUN GOLD

One interesting example of the possible complications of employing dehybridization to create a stable, open-pollinated version is in the case of Sun Gold, the widely popular orange hybrid cherry tomato. The vast majority of people who taste Sun Gold proclaim it as one of their favorites. Yet developing a passion for F_1 hybrid tomatoes is risky. If for some reason the company decides that it no longer wishes to sell the variety, all Sun Gold tomato addicts are out of luck. The only hope is that efforts in dehybridization result in a very good approximation of Sun Gold, if not a nearly exact replica.

Years and years of work and countless efforts sprinkled around the world seem to have been (and continue to be) expended on this little project. I actually tried my hand at this in the early 1990s. I saved seed from a few Sun Gold fruit and grew out half a dozen or so of the F_2 generation in the following seasons. One thing I noticed immediately was the diversity of results, indicating the complexity of the cross or crosses involved. I had regular- and potato-leaf plants and tomatoes that were red, yellow, or gold, on the larger or small side, and some with blossom end points. The flavor of all of them in general was quite good, but none of them approached the intensity and just plain yummy flavor of Sun Gold itself. In flipping through the SSE yearbook over the years I've noticed Sun Gold Select, Sun Gold Select II, Big Sun Gold, and Big Sun Gold Select showing up as offspring of the hybrid. I've tried several and read through descriptions, and though some are quite good, none of them quite match the unrivaled excellence of Sun Gold.

MEET THE VARIETIES
NEW DWARFS

Dwarf Mr. Snow

The story of Dwarf Mr. Snow begins in 2006 in Australia when my dwarf project co-leader, Patrina Nuske Small, made a cross between Golden Dwarf Champion and Green Giant. Golden Dwarf Champion is a true American heirloom, having been developed and released by the Burpee Seed Company in the 1890s. It is also one of the few authentic, established dwarf-growing tomato varieties that existed before our project. Green Giant is a variety that came to prominence only recently, having been discovered by Reinhard Kraft (a gardener in Germany) as a variant of Lillian's Yellow Heirloom. I received it for trial in 2004.

Patrina named the cross Sneezy, and it was one of the original nine crosses she carried out in 2006 to kick off the Dwarf Tomato Breeding Project. I grew out the Sneezy F_1 hybrid and sent seeds from the delicious yellow tomato back to Australia. Another team member there, David Lockwood, grew out a number of the seeds, and one of the potato-leaf dwarf varieties he obtained produced delicious, good-sized yellow tomatoes, which he named Summertime Gold 4. Seeds from this selection were returned to me, and I grew out four plants in 2007. One of them produced truly wonderfully delicious medium-to-large, ivory-colored fruits on a potato-leaf dwarf-growing plant. Clearly, this new tomato deserved a name.

Now, back in the early 1990s, an editorial writer for the *Raleigh News and Observer* named A. C. Snow wrote a column on tomatoes. In it, he bemoaned the state of affairs of the tomatoes he was purchasing at local markets, noting that they had nothing in common with the delicious tomatoes he ate as a youth. I sent Mr. Snow an e-mail offering a visit to my Raleigh garden to see if I could demonstrate that the flavors he longed for did, in fact, still exist. Visit he did, and upon tasting a perfectly ripe Brandywine, as well as a few others, he was moved to write a column about his visit to my garden, and his renewed faith in tomatoes. I have maintained occasional contact with Mr. Snow through the years and always enjoy his musings on daily life in the South.

Knowing how much Mr. Snow loved tomatoes and respected my efforts to spread the word about heirlooms and other great tomatoes, I decided on Dwarf Mr. Snow. It seemed fitting, also, because the tomato we were trying to name was as white as snow. After a few more years of selection and work, it was released to the public through the Victory Seed Company in 2010. As far as I can tell, people really love it, and I am convinced that a few generations hence, it will be a favored heirloom tomato indeed. And, after Mr. Snow found out that a tomato was named after him, he expressed his delight in one of his weekly editorial columns. It was very satisfying to bring joy to someone, especially in the form of immortalization with a great tomato!

Part of the fun of the Dwarf Tomato Breeding Project is the rule that whoever discovers a unique, great tomato variety from the project gets to provide the name. The same year that I discovered and named Dwarf Mr. Snow, I found other great varieties. Joining A. C. Snow in tomato-name immortality are my wife, Susan (Dwarf Sweet Sue); my daughter Sara (Sarandipity); my daughter Caitlin (Dwarf Caitydid); my dad (Dwarf Wild Fred); and my wife's uncle (Dwarf Kelly Green). What fun it is to name tomatoes, although it's not as easy as you would think.

Rosella Purple

This is another of our Dwarf Project tomatoes, named by project co-leader Patrina Nuske Small. In the line that led to Rosella Purple are some real tomato genetic mysteries. It all started when Patrina crossed a wonderful, large-fruited, pink, potato-leaf variety called Stump of the World (a variety popularized by Ben Quisenberry of Brandywine fame) with a medium- to small-fruited, unexceptional but compact dwarf variety called Budai Torpe. She named this cross Sleepy, and it is one of the original nine crosses that started the Dwarf Tomato Breeding Project. Working with Sleepy, Patrina named one of her dwarf finds in the F_3 generation Rosella Crimson, reversing the name of the lovely Australian parrot known as the crimson rosella.

The compact dwarf plant produced medium-size delicious pink tomatoes, worthy of being a new variety. Much to our surprise, though, one of the plants that I grew from Patrina's seed unexpectedly produced medium-size purple (not pink!) tomatoes, which I named Rosella Purple. (In fact, work on Rosella Crimson continues; it has proven to be one of the more challenging of the dwarf varieties to stabilize.) Patrina and I, assisted by many others on the project, traded seeds back and forth and worked on selecting Rosella Purple to a reproducibly stable state, and we were delighted to note that the flavor actually improved as our work proceeded. The big mystery, of course, is of the source of the purple color, since the parents of the original cross were red and pink. Clearly something in this line allowed a recessive trait to show itself.

The fruit of Rosella Purple, in the half-pound range, are produced in abundance on a relatively compact dwarf plant that tops out at 3 to 4 feet tall. The skin is clear, and the flesh is the dark crimson with deeper tones characteristic of the "black" tomatoes. Rosella Purple is therefore a "pink/black," or purple tomato, similar in color to Cherokee Purple or Black from Tula. The flavor is truly outstanding. The ability to get such flavor on a tomato plant easily grown in a 5-gallon pot on a porch or deck is a true advance in the selection of tomatoes possible for space-constrained growers. Rosella Purple was introduced to the public in 2010 by Victory Seeds.

8

Q & A

ANY READER WILL FIND AMONG the pages of gardening books, magazines, and websites dozens and dozens of suggestions, tips, and hints that may sound rather unbelievable. I like to call these gardening urban legends, often unproven but, for whatever reason, adopted by this or that gardener and then passed along as folklore during discussions with friends and neighbors. I'm sure many great ideas are included, as well as procedures that are most effective in a specific locale, but some of these "tips" are just money-making scams. The information age has done both great and awful things for our knowledge. In gardening (as with money-making schemes), if something seems too good to be true, it probably is. I've found through the years that much of what works in growing tomatoes is common sense and easily reasoned.

This chapter consists of a list of various tomato-related myths, urban legends, and off-beat theories that I've collected through the years, along with my experience with or assessment of the approach. But I'm also still learning all the time, so read them, process them, and reason my assertions, and then make up your own mind.

Why not try testing some of these theories yourself? If you do, remember to carefully control the experiment. Try to use the same variety of tomato in the same type of garden environment in close proximity, to ensure that the only thing you're doing differently to one of the plants is the practice you're testing; in other words, introduce only one change, or variable, and treat each plant the same throughout the growing season. I've often read about the placebo effect in play as the cause for the noted improvement — if you really believe something should work, the desire to see the effect overcomes objective evaluation. In some cases, the tested plant may be just slightly more coddled or cared for, and it becomes that little bit of extra care that makes it seem like something is working, when it really is not.

Is it risky to grow non-hybrid/heirloom tomato varieties? Will they get diseased? Is it safer to stick with hybrid varieties?

Not necessarily. It depends on your goals. In general, many hybrid tomatoes were bred to handle various diseases (see page 175). Flavor was often not the primary goal, setting up what seems to be a trade-off between yield and flavor. The response to this question has many dependencies, all related on how and where you garden and your particular goals (variety, yield, flavor, or seed saving as examples).

I was so interested in this topic of hybrid versus heirloom that I undertook a detailed study, the results of which can be found on page 228. My conclusion was that the heirloom varieties I grew equaled, and in many cases exceeded, the performance of most of the hybrids. What is most important is to understand the goal of your garden, try different varieties and approaches, and make future changes based on what you experience.

Is the "Tomato Potato Plant" I saw advertised in the Sunday paper for real? Sounds like a great way to get both from the same plant.

If it sounds too good to be true, it probably is. I've seen Sunday newspaper inserts through the years offering a plant that has tomatoes above and potatoes below, or bush-type tomato plants capable of immense fruit or 100-pound yields. My experience tells me that it is best to avoid such offers, or at least to approach them as curiosities only if you are willing to part with the money it takes to purchase them.

Since potatoes and tomatoes are both from the Solanaceae family, it is possible to graft a tomato "top" onto a potato stem. (I've also read of rooting tomato stems inside potatoes.) But there are so many potential issues with this that it's hard to see why a person would want to dedicate valuable garden space to the endeavor. The potato gets the energy needed to produce potatoes by producing rampant vines. If a tomato plant top is grafted onto a potato plant stem, one root and vascular system is now responsible for the nutrition needed to produce two crops — the potatoes below the soil, and the tomatoes on the top growth. I'd love to hear the results from anyone who gives it a try and compares yields of potatoes and tomatoes of the same varieties from normally grown plants.

To make matters worse, potatoes often produce seed balls that look like small tomatoes and are very toxic — mistakenly eating one, thinking it was a cherry tomato, would be a very bad thing to do for your physical well-being. Finally, potato plants are magnets for the potato beetle. The beetles would be just as happy to munch on the close-by tomato foliage as on the harder-to-find potato foliage.

It is hard to keep up with all of the overhyped gardening merchandise, typically advertised in Sunday-paper inserts (which seems to be where they first appeared many years ago) and now also found in popup Internet ads or on garden blogs and websites. Exaggerations of fruit size and plant yields are nothing new, as a look through old seed catalogs (dating into the mid-1800s) confirms.

Can I sow tomato seeds outdoors in the winter? I've heard that this is a great way to get an earlier harvest with less work.

It depends on the length of your growing season and which varieties you choose to grow. Winter sowing is the process of planting tomato seeds outdoors in containers at about the same time you would plant them indoors. The concept is that the weather experienced between planting and spring will provide well hardened-off, hardy seedlings, without the need to purchase and set up the materials used for indoor seed starting. In a way, many of us observe winter sowing in action when we notice that seedlings sprout in the spring in the containers or garden areas from last year. Perhaps some tomatoes or peppers fell off the vine and rotted on the ground, and when the conditions warmed and became appropriate in the spring, there was germination. Many of my seedling customers return to me to tell tales of seedlings of tomatoes popping up throughout their gardens, and

they wonder what they are and whether they will grow. Winter sowing is a way to control that process so that you are confident of the varieties that pop up. If you are unable to start your own tomato seed indoors, winter sowing may be a good solution for you, given a long enough growing season.

Direct seeding is a variation on winter sowing; the timing and location are different. With winter sowing, you begin when it is still essentially winter and create a somewhat protected location into which you do all of your sowing in various containers. Direct spring sowing is planting seed in place where you will grow the plants to maturity. This certainly is a method that can work in long-season areas, or to extend the season; the seeds will germinate, and as long as you don't have a late and unexpected frost, they will grow on, become hardened off naturally, and produce full-size plants, but only where the season will allow before the killing frost; you are essentially cutting a few months from the plant's life cycle, since the two months from seed to a plant ready to be set out will be happening in the garden. Each year, it seems that a few tomato seedlings from dropped fruit escape my detection and produce surprise plants, though it seems to be most common with cherry tomatoes. The variety Mexico Midget appears here and there throughout our yard each season, much to the delight of our chocolate lab, Buddy!

If I grow tomatoes next to each other, will they cross-pollinate and produce fruits that will appear as a cross of the varieties?

No. A tomato plant of a given variety will produce that variety, even if some of the flowers get crossed by bees with pollen from another variety that season. Seeds saved from such crossed tomatoes, however, will produce hybrid plants the following season, which will likely appear to be very different.

If I separate tomato varieties in my garden by a large distance, will that guarantee that the tomatoes will not cross-pollinate?

Separation helps but is not a guarantee; it depends on what the bees do. Tomatoes are one of the best plants for seed savers because they tend to self-pollinate the vast majority of the time — over 90 percent. Even if curious bees visit open flowers, the pollination deed has likely already been done. But likely isn't always; consider the time it takes a bee to move from one corner of a 100-foot-long garden to the other, and you will understand that separation doesn't provide a guarantee of purity. Growing other flowering plants between varieties will reduce chances of crossing.

The only way to truly guarantee that tomatoes will not be crossed by bees is to use a physical barrier to isolate flowers. You could try securing a homemade sack of a very light, air-permeable material such as row-cover fabric around an unopened blossom cluster, or use a cage that is wrapped in screening or row cover fabric. I plant my tomatoes very close together and grow out thousands of seedlings each year, and find that by saving seeds from the earliest

fruit (whose flowers would have been self-pollinated early, before the bees took much of an interest in the blossoms), my seed purity is greater than 98 percent. If you choose to not use a physical barrier but want to save seeds, be aware of the bee population in your garden. Save seeds from fruit that started to develop before the bees' arrival, or wait until late in the season, when bees find other types of flowers to keep them busy.

Will setting up a fan to blow on young tomato seedlings cause them to develop sturdy stems and end up hardier?

It can't hurt. Air circulation certainly benefits the health of very young tomato seedlings by reducing the incidence of fungal diseases caused by stagnant, humid conditions. Toughening up the plants by aiming a fan at them is often mentioned. And, in fact, in a study done at Cornell University, lightly brushing tomato seedlings (with a broomstick or plastic pipe) was shown to slow the rate of stem elongation, leading to sturdier, stockier plants.

Personally, I've always felt that nature provides the perfect solution: a nice, gentle spring breeze. But plugging in a fan certainly can't hurt either.

Will sticking a copper wire through a tomato plant stimulate the plant defense mechanism through an electric charge?

No. The premise is that either through movement of copper ions into the plant vascular system or through some effect of a charge built up through the copper wire, some protective properties are transmitted to the growing tomato plant so that it is less likely to suffer from disease. The procedure involves passing a copper wire through the lower stem of a young tomato plant and passing part of the wire into the ground. I've also found variations that include spraying the copper wire–treated plants with 3 percent hydrogen peroxide. As a scientist, I can't come up with a basis for the effectiveness of this treatment. Most people who tried this reported no difference, and as with all of these types of unique or unusual remedies, a very few report high effectiveness.

Will injuring some tomato foliage early on cause the plant to create its own disease-fighting agents and help the plant be healthier?

No. The idea is that crinkling or otherwise maiming some young tomato foliage will cause the plant to fire up its defense mechanisms, which would hopefully lead to being able to fight any disease that comes to infect the plant. Unfortunately, there is no sound scientific basis for such a phenomenon. Indeed, intentional injury to tomato foliage produces an easy entryway for disease.

Will pruned tomato plants produce larger fruit than unpruned plants?

In my experience, no; the size of a tomato is controlled by its specific genes. Any variety will grow as large as it can when it's given optimum growing conditions.

Will removing all of the foliage from the plant increase the production, flavor, or quality of the tomatoes?

No. In fact, this is detrimental. The removal of all foliage from tomato plants is an invitation to numerous issues and will significantly reduce the quality of whatever fruits manage to grow to maturity. With no foliage to shield the tomatoes from the searing heat of summer's direct sun, sunscald would have a significant impact on most, if not all, of the tomatoes, ruining the quality. Tomato flavors develop because of photosynthesis going on in the leaves. One of the reasons that indeterminate tomato varieties

are generally more intensely flavored than determinate varieties is the vastly increased leaf area, which allows far more chemistry to go on in the plants and results in increased flavor. This is thought to be the reasoning for the generally superior flavor of indeterminate tomato varieties when compared to determinate types.

Will using leaf or grass mulch around the base of the tomato plant allow for the formation of substances that fight some of the various tomato afflictions, leading to healthier plants?

No. As a physical barrier that prevents spores from potentially diseased soil from splashing up onto the lower foliage of the plants, mulch is a very effective tool that contributes to a healthy garden. Mulching also helps to slow down water loss from the soil on very hot days, which can help reduce the formation of blossom end rot. As mulch breaks down it forms humus, which acts as a natural fertilizer for the plant. But there is no direct evidence that mulch itself leads to the formation or existence of actual disease-fighting or disease-blocking compounds.

Will allowing my tomatoes to sprawl, uncaged and unstaked, result in greater yields?

In theory, yes, but many factors will negate the advantages. Sprawling tomatoes are certainly easier to manage, and it makes sense that just letting them go will allow them to produce enormously. There are many caveats, however. Most soils contain various bacterial, viral, or fungal agents that can harm the tomato plant and fruit in various ways, so yield is easily compromised by disease, if and when it hits. A good layer of a barrier substance, such as straw, leaf mulch, or non-treated grass clippings, is necessary to keep the soil off the tomato foliage and fruit. Even still, mice, voles, and slugs can do great damage to the tomatoes. A tomato cage, allowing vertical, cleaner growth, achieves the same sort of yield with less potential loss from disease. Sprawling plants are also a potential logistical problem in terms of getting in to harvest the tomatoes without stepping on and injuring the plant. Varieties can also get easily confused, which is detrimental to accurate seed saving. Indeterminate tomato vines can easily reach 10 feet or more in a long growing season, and all of those side shoots (a.k.a. suckers) will reach nearly as long. This explains both the incredible yield potential and the tangle of tomato plant that will likely result.

Do those Topsy Turvy containers really work?

This novelty works reasonably well, but only for small-fruited tomatoes, with frequent watering. The Topsy Turvy is a name brand of a widely advertised hanging planter with a simple premise — the plants are grown upside down, with the seedling emerging from the bottom, roots upward, watered from the top, and hung from a secure hook. This can be considered a space-saving and unique way to grow plants, and it does work.

Strawberries, which produce numerous runners, may do quite well in such a container. Training flowers such as some petunia varieties would be lovely in one. But it is very limited in its use for eggplant because of the upright nature of the growth and vigor of the plant, and it is essentially useless for the fragile-stemmed sweet peppers, though it actually works to some degree for tomatoes, with two significant caveats: A Topsy Turvy device doesn't hold much soil, so it is limited to one tomato plant for any hope of a decent yield. And because larger tomatoes become heavy as they ripen, a single cherry tomato plant is probably the best use. I planted a single Sun Gold plant in a Topsy Turvy a few years ago and hung it off my deck. As long as I kept it well watered, it thrived, and it hung 8 feet down to the ground, producing a decent crop of tomatoes along the way. Our dogs, in particular, thought it was a splendid idea, especially when tomatoes started to appear at ground level.

Will leaving my grow lights on for 24 hours a day result in healthier tomato transplants?

No. In fact, this practice can be detrimental. "More should be better" is a common supposition. If tomato seedlings benefit from grow lights, then the longer they are on, the happier the seedlings? Well, no. Keep in mind that plants in nature receive a break in the action between sunset and sunrise. Research indicates that the optimum amount of time for tomato seedlings to be under light is 14 hours. No additional benefits are seen between 14 and 20 hours, and possible detrimental effects are seen beyond that (the theory is that a buildup of excess starch and sugars in the foliage promoted by the nonstop light leads to foliage damage).

I turn my grow lights on when I wake up and turn them off when I go to bed, which provides light between 8 a.m. and 10 p.m.; this has worked fine for me. It isn't necessary to be absolutely specific about the start and end times, so you can fit this to your own particular morning and evening activities. A timer can be very helpful if your schedule is not predictable.

Is transplanting my tomatoes really necessary?

Transplanting is strongly recommended. Transplanting seems to be the norm for the gardeners who start their own tomato seeds, though it is true that, in some climates and for early varieties in particular, direct seeding can work. Often, we find in our garden volunteer plants from the previous season's dropped tomatoes that emerge once the soil warms and, if we leave them be, turn into fruiting plants. This then raises questions of whether one needs to transplant.

For most of us, transplanting is a necessity because volunteer- or direct-seeded seedlings wouldn't nearly match the yields of those plants that get their start indoors. In addition, transplanting allows for very vigorous and healthy seedlings, due to root development along the part of the stem that is buried during the transplanting step (see page 64).

Will my garden be more successful if I plant by the phases of the moon?

Evidence of success are anecdotal, and those who abide by it are passionate. I've not tried it myself. Planting governed by the phases of the moon seems to stem from a long-held belief that the moon controls moisture, a principle dating back to Pliny the Elder in the first century C.E. How this theory extrapolates to when we do specific tasks today depends on the beliefs of the individual gardener. The guidelines indicate that the various phases of the moon will be beneficial to different gardening tasks, whether it be seed sowing, planting, or weeding. As always with suggestions that seem to be more legend or myth-based than science-based, these provide great opportunities for small projects so that you can explore for yourself which ones do or don't work for you.

If I save seeds from the first tomatoes that ripen on a particular plant, and I do this for several seasons, will I end up with a tomato that is earlier than the one I started with?

No, since the genetic material in all seeds from all tomatoes on an uncrossed, open-pollinated plant will be the same.

If I save seeds from tomatoes with blossom end rot (BER), will all of the tomato plants from the saved seeds produce tomatoes with that affliction?

No. If a tomato variety is open pollinated, all of the genetic material in each seed in each fruit on a plant is the same. A physiological issue like BER is a result of the conditions experienced during the formation or ripening of the fruit, not of anything genetically distinct about the particular affected tomato, so seed saved from a fruit with BER should be fine. The only exception would be if every single fruit on the plant, and other plants grown from the source seed, suffers from BER; then it is possible that something genetic about the variety is making it prone to developing BER.

Should I remove all suckers from a tomato plant? I've heard that they sap energy from the plant.

No, they don't sap energy. In many cases, leaving some or all can be beneficial. It was actually the suckering topic that gave me the idea of dedicating this chapter to tomato myths and urban legends, because I get more questions about sucker removal than just about anything else during our spring seedling sales. I cover side shoots/suckers in depth on page 92. The short answer is that suckers are simply additional growing stems and do not sap energy from the plant. Indeed, there are more reasons to keep them than to prune them.

Will shading my tomatoes to reduce the temperature or using calcium spray on the blossoms prevent blossom drop and increase the chances of a good crop?

These practices have limited effectiveness at best. Each particular tomato variety has a temperature and humidity range that facilitates pollination of the flowers and therefore fruit set. Tomatoes from cherry to medium size seem to be far less fussy and yield more reliably over a wide range of conditions. The large, irregularly shaped beefsteak tomatoes can be fussier, though it is variety specific. Tomato flowers should, in theory, pollinate as they open, from the brushing of the pollen-releasing anthers against the receptive tip of the pistil. Though it is impossible to control the humidity experienced by a tomato blossom growing outdoors, screening on a very hot day could provide a microclimate sufficient to increase the probability of pollination. To the tomato grower who is salivating at the thought of the crop to come, any simple efforts such as these are worth a chance.

The various hormone and calcium sprays that are applied directly to the blossoms are another story entirely. They are intended for use primarily to promote pollination and fruit set in cooler climates, and their effectiveness in the more problematic hot, humid conditions is likely very low. And even in those cases where fruit set succeeds, the resulting tomatoes tend to be seedless (parthenocarpy), resulting in unpleasantly mealy, even bland fruit that don't approach the quality of those that pollinate naturally.

My tomatoes won't set fruit! Is that because there are no honeybees?

No. I hear this comment frequently each season during our seedling sales. Many of my plant customers find that their tomato yields are lower than they had hoped, and since many yards are short of honeybees it can seem like a correlation. First, honeybees are not particularly attracted to tomato flowers (many other bee types are, though). Second, tomatoes are self-pollinating, and the deed is typically done as the flower opens, meaning bees are not needed. More important is a much-overlooked fact: the often unrelenting heat and humidity of the summers, particularly over the last decade. Lack of tomatoes is a result of excessive temperature or humidity experienced during the flowering of the particular variety.

I have a lush, healthy plant with no fruit at all. What's wrong with it?

Most likely the blossoms dropped before fruit set. When a tomato plant is healthy but fruitless, three explanations are possible. If absolutely no blossoms are produced throughout the entire season, the plant is a "mule," possessing a rare genetic mutation. In all the years I've grown tomatoes, I've observed this only a couple of times. Far more common is the impact of weather conditions. Perhaps the plant produces flowers regularly along the vine, as it should, but the flowers shrivel and drop off, leaving no tomatoes at all. This phenomenon is quite common during very hot, humid periods with the large beefsteak-type varieties (it is highly unlikely to happen to small- or medium-fruited varieties), particularly if the plants are heavily pruned, leaving only the main growing stem. Since the numbers of flower clusters will be limited, if the conditions favored for pollination for many of the clusters are unsuitable, the result could be 8 feet or more of very little harvest. The best way to avoid this issue if you want to grow large-fruited tomatoes in a hot climate is to hedge your bets on the weather: cage or only minimally prune the plants and/or stagger the tomato plantings so that the varieties you desire are flowering throughout the growing season.

I've heard reports that overly fertilizing tomato plants (especially with an excess of nitrogen) will actually lead to an excess of vine and a relative lack of tomatoes, but I've never observed this in my gardens. The goal of a tomato plant is to create seed, so that it carries on from generation to generation; the flowers, and subsequent tomatoes, should happen, given appropriate weather and plant nutrition.

Will getting water on the tomato plant foliage lead to disease or damage from the sun's hot rays on the beads of water?

Damage from the sun is unlikely, but promotion of disease is possible, especially if soil gets splashed on lower foliage. Rain in cool weather can also lead to late blight, if the disease agent is present. Many tomato gardeners do everything they can to prevent wet tomato foliage, but consider what happens to the plant in a heavy rainstorm. It isn't so much about wet foliage as it is about what may be splashing on the plant that could cause harm. Since many tomato diseases are embedded in garden soil, the most important thing is to prevent soil from splashing up onto the lower foliage by mulching well around the base of the plant. Watering large tomato plants from above is potentially wasteful, as the water cascades off the foliage but, especially if the plant is large, may not provide the deep watering that the plants require. I've experienced no issues at all with watering flats of tomato seedlings from above using a gentle spray. But once a plant is seated in its final location, be it pot or garden, it is far more effective to water around the base of the plant to allow for a good soaking of the root zone.

Is it true that the tomatoes on my plant won't ripen if they're hidden by the leaves?

No. Tomatoes ripen when they are good and ready, informed by the particular genetics of a given variety. Weather will certainly play a role, but the location of the tomato on the plant doesn't impact its ripening time at all. In fact, exposed green tomatoes that become roasted by direct sun in the middle of the summer are prone to sunscald. The main problems with deeply hidden tomatoes are the propensity to actually miss them when they are ripe and the opportunity for increased pest attacks because they are more difficult to monitor.

Do I need to vibrate my tomato blossoms or shake the plants each day to get good pollination?

This isn't necessary, but it could help if the plants are grown to maturity in greenhouses.

The principle behind shaking or vibrating tomato blossoms is that the flowers may need a bit of help to ensure that the pollen from the anther is captured by the tip of the pistil to ensure fruit formation. This is a common practice with greenhouse tomato growers, since there are no natural breezes to move the flowers. It shouldn't be necessary to vibrate tomato plants grown outdoors.

To test the effectiveness of this practice, it would be interesting to have the same tomato variety side by side, and to vibrate each flower daily (or shake the entire plant — gently, of course) on one plant and totally ignore the other plant. Be sure to collect information on numbers of tomatoes set and total fruit weight. If you do this, I'd love to know how the experiment turns out.

Should I pinch all blossoms from my transplants, so that the energy goes into the plant, not the first tomatoes?

No. The supposition of this garden myth is that if you have transplants that possess buds or open flowers, or even small set fruit, all of these must be removed when the transplant is set into its final location so that it puts its effort into establishing the plant, not dealing with maintenance of the flowers or tiny fruit. My personal experience is that the plant knows best, and I've often gotten some nice early tomatoes (far in advance of posted maturity dates) by leaving any flowers or fruit as I settle the transplant into its resting place. I've observed no delay or setbacks with the establishment of the plant; grown side by side, there is no difference in the health or size, over time, of plants with or without flowers and fruit when planted. In my opinion, all you are doing by plucking off those first flowers and

fruit is delaying gratification. But, as in the case of the myths that are easily tested in the garden, it is better to "do the experiment" and find out for yourself.

Are tomato diseases passed along on seeds?

Sometimes. Unfortunately for tomato seed savers, a number of serious tomato diseases can indeed be carried on saved seed. The three general types of tomato diseases, caused by viral, bacterial, or fungal agents, are covered in detail in chapter 9. Even though fungal pathogens are the greatest in number, they tend to be more of a plant surface issue and, though they can be harbored on seed coats, are far less of an issue. Bacteria and viruses are better able to navigate the vascular system of the plant and wind up in the seed embryo, and hence the final seeds. Simple seed treatments can be used to minimize the pass-along disease potential from infected seeds. Of course, the best prevention is to ensure seed is saved from healthy plants only.

If I save seeds from hybrid tomatoes, will they germinate?

Yes. Any properly saved tomato seed will germinate, whether saved from open-pollinated or hybrid varieties. So it is not *if* you get a tomato plant; it becomes all about *what* you get. The F_2 generation, which is what you are working with if you plant such seeds, will segregate into plants that genetically express the parents of the hybrid, depending on which characteristics are dominant or recessive. In summary, all bets are off when you grow tomato plants from seeds saved from hybrids. You may like what you find . . . and you may not!

Do I need to plant at least two of each type of tomato so that they cross-pollinate and produce tomatoes?

No. I am asked this question often during seedling sales by gardeners with limited space. They want to try out as many varieties as possible but think that two of each is needed to get a decent crop. Fortunately for the intrepid tomato explorer, tomato flowers are "perfect," meaning that they effectively pollinate themselves upon opening. If you have room for 12 tomato plants in your garden, and you love variety, select 12 different tomato varieties and enjoy the experience of discovery as you learn about each of them.

Is it true that pink or yellow varieties are less acidic than red varieties?

No. According to a study carried out by the USDA in 1977, just about all tomatoes are similarly acidic.

Relative acidity is measured on a pH scale of 0 to 14, with 0 being extremely acidic and 14 being extremely alkaline. Acidity and sweetness are not opposites, however. In fact, honey (with a pH of around 3.8) is actually more acidic than tomatoes (which have a pH range of 4–4.6). Some tomatoes taste sweeter than others simply

because their sugar content is higher, which masks the acidity.

When considering which tomato varieties to can, feel free to choose anything that is in good edible ripe condition. Acidity does drop off a bit with age, so *very* ripe tomatoes, which often taste unpleasantly sweet, are likely have elevated pH levels and are best kept out of the canning mix; sauce may be the perfect use for those that have been neglected on the counter or vine just a bit too long.

If I grow a tomato variety for enough years in my garden, and I save seeds each year, will it become genetically more adapted to my climate over the years?

No. A particular tomato variety has a specific set of genes that control how it grows, how it looks, and how it tastes. If, over a period of time, the tomato seems to change and adapt, it is likely due to either subtle changes in how it is grown, how the weather affects its growth, or even an inadvertent selection for different characteristics by the seed saver. For example, if half a dozen plants of a particular variety are grown each season and one plant does slightly better, and the variety is carried on by saving seeds from tomatoes only from the superior plant, something is likely genetically different about that plant. So, in this case, it isn't that the genes are changing or adapting. Rather, a genetically slightly different tomato was found and is now the basis for refinement into a new variety.

Acidity of Tomato Varieties

pH	
	RED
4.37	Earliana
4.37	Valiant
4.29	Rutgers
4.26	Fantastic
4.14	Early Girl
	PINK
4.37	Ponderosa
4.3	Arkansas Traveler
4.3	Oxheart
	YELLOW
4.40	Yellow Pear
4.22	Jubilee
4.21	Sunray
	WHITE
4.3	White Beauty
4.21	White Queen
4.16	Snowball
	CHERRY
4.4	Red Cherry
4.3	Basket Pak
4.15	Small Fry

9

Troubleshooting DISEASES, PESTS, AND Other Problems

EVERY GARDENER WHO HAS EVER attempted to grow a tomato plant knows that, in many areas of the country, the assortment of challenges to a healthy plant and delicious tomatoes can be daunting. Indeed, my view is that tomatoes provide a challenge that isn't so different from that of healthy, bountiful roses. It seems as though every disease, pest, and weather hiccup will affect the tomato crop. And for every tomato-growing problem, there are myriad solutions — some tried and true, some new and advertised to be utter miracles.

The pest, critter, and physiological problems described in the sections below are far easier nuts to crack than identifying and dealing with the many dreaded tomato diseases. For one thing, it's actually possible to do something positive and helpful if the problem is related to pests or growing conditions. Erecting a fence, moving a plant, mulching it more completely, plucking off a hungry worm — these activities may not be much fun, but the effort is rewarded with a good chance of the eagerly anticipated tomato harvest.

Any one of these problems can lead to high anxiety and nightmares for avid tomato growers, but the effects on your tomato-growing efforts can be minimized if you understand how they arise and take appropriate control measures. The best way to learn about solving tomato problems is to contact your local Extension agent; there is no substitute for local expertise to help guide you through your specific situation.

COMMON PHYSIOLOGICAL PROBLEMS

Sometimes your tomatoes struggle because a disease has entered the plant, and sometimes a critter of some sort takes interest in your efforts. Physiological tomato problems, though, fit into neither category (though attack by diseases or critters can certainly lead to some of them); instead, these issues are related to what the plant is experiencing in its physical environment. I will share my coping methods, my successes, and my continued failures, but those who garden have their own solutions that work best for them. Consider this a starting point for your own research and experiments.

Learning to "read" your plants is a useful skill to develop. For example, a wilting plant tells many possible stories. On a blazing hot afternoon, healthy tomato plants conserve moisture by curling up their foliage; as the sun sets, such plants return to normal. A well-watered tomato plant that has significant wilt throughout the day is an indication that something is amiss. Wilting that is accompanied by a significant change in foliage color — from shades of yellow deepening to brown or even black — is a serious sign of disease; see page 175 for more information on tomato diseases.

Observation is key when it comes to foliage colors, too. After growing and carefully observing more than 1,000 varieties of tomatoes, I've been amazed at the variations of the color green that exist. Large bicolored beefsteaks, as well as many of the compact rugose-leaved dwarfs, have foliage with a luxuriant, nearly bluish-green hue. Some of the large pink beefsteak types have a more medium to pale green hue, and a few interesting mutations have chartreuse or even yellow foliage, even when completely healthy. And tomato foliage on a given plant and variety does vary throughout the growing season, from the purple-tinged foliage of youthful seedlings under grow lights in a cool garage to the nearly brilliant yellowish-green foliage of a plant that adjusts to its new home after transplanting and is starting to spread its roots and stretch its stem upward toward the warmth of the sun. Once again, it's helpful to learn to read your garden, so that you can distinguish between the normal evolutions of colors throughout the season and an unexpected, eventually unwanted, color change that indicates trouble is afoot.

INSECT PESTS

We gardeners are well aware that planting a garden is like laying out a welcome mat to a host of critters as attracted to our efforts as we are. The strategies for minimizing pest problems allow a chance for the gardener to better understand the various non-human "neighbors" of all types that want to share in the delight of the garden. Each gardener must also make various philosophical choices — spanning from live and let live to reasonable sacrifice to total annihilation of the offending agents, and, similarly, between the various natural/organic agents to chemicals approaches.

My own battles with insects vary widely from season to season. I was troubled by whitefly only once, while aphids visit me annually. However, they tend to be a problem only early in the season, on young seedlings, and tend to affect young pepper plants more often than young tomatoes.

Hornworms vary from abundant (rarely) to absent (occasionally). My own approach for dealing with insect pests is to observe, locate, pick, and destroy. If I find green tomatoes with the telltale entry holes of fruitworm, I remove and destroy the affected fruit. In all my years of gardening, the need to resort to insecticidal sprays has been rare, and for that I consider myself fortunate.

Natural Pest Control Options

BENEFICIAL INSECT LARVAE — The larvae of green lacewings, lady beetles, assassin bugs, and praying mantis are considered highly beneficial to have in your garden, as they feed on many different garden pests that may be causing you tomato troubles.

***BACILLUS THURINGIENSIS* (BT)** — Bt is a soil-dwelling bacteria that is an effective organic pesticide.

DIATOMACEOUS EARTH (DE) — This powdery white substance is composed of the fossilized remains of tiny diatoms, which are a hard-shelled algae. It is effective as an insecticide for soft-bodied pests, because of its ability to absorb lipids from the exoskeleton, leading to dehydration and death.

PERMATILL — This coarse rocky material can be added to the soil around the roots of plants to dissuade attacks from voles.

ROTENONE — A naturally occurring compound in the stems of a number of seeds and plants, such as jicama, rotenone is used as a pesticide and insecticide.

INSECTICIDAL SOAP — These solutions contain potassium salts of fatty acids and work best as insecticides for soft-bodied pests.

PYRETHRUM — This natural compound is derived from chrysanthemum and has been long used as an organic insecticide.

MAMMALS AND BIRDS

The fun and excitement of gardening loses steam in a hurry when you start doing battle with critters. Whether you're losing valued seedlings or seeing the telltale bites taken from tomatoes nearing peak ripeness, the damage sends gardeners rapidly around the grief cycle of disbelief, disappointment, sadness, and anger. Acceptance is not a part of the equation if there is to be a useful harvest, so it becomes a war to be won. Most critter attacks will vary from season to season, depending on population and alternate food sources. But once they've targeted your garden as a great place to eat, the critters may have a long memory.

Whether the opponent is of the short or tall four-legged variety or one that can swoop and perch, deterrent options fall into just a few categories.

Physical Barriers

WRAPPED CAGES — One option for preventing ground squirrel, rabbit, or bird damage is to cage the tomato and then wrap a mesh-type fencing around the cage, ensuring that the mesh holes are far smaller than the critter. Another possible method is to loosely wrap the still-green tomatoes with a very light material such as floating row cover or nylon, and keep the covering on until the tomatoes are ripe and ready to pick. The downside is that these efforts can become costly and present access issues when you need to work with your plants throughout the growing season.

FENCING — The best solution for those who are plagued with rabbit or groundhog attacks is a fenced garden. Rabbits have been reported to jump short fences, so the height of the fence should be carefully considered. Mesh size is also critical. Both groundhogs and rabbits

will dig under a fence, so it's best to bury the bottom of the fence at least a foot below the ground. We have had great success using a ScareCrow motion-activated water sprayer to deter deer (see page 177), but if deer pressure in your area is significant, a tall multi-strand electric fence may be your only option.

"Live Dissuaders"

Those of us who garden in suburban or rural areas are aware of the wide variety of creatures interacting in and around our yards. For every pest that is attracted to the tomatoes we grow, other creatures are present to provide natural deterrents. For that most stubborn and destructive four-legged pest, deer, the family dog is really the only living aid in most areas. But other natural inhabitants, such as hawks, snakes, and cats, could be quite helpful in dealing with birds and smaller mammals. Creating an environment to ensure the greatest living natural diversity not only makes working in the garden more interesting but can lend a helping hand to ensuring that the tomatoes you grow end up in your harvest baskets and not in the mouths of wild animals. Of course, be sure to know your snakes, so you can ignore those that are beneficial and avoid those that are dangerous.

Repellent Materials or Devices

There is no shortage of substances and devices touted to dissuade hungry, uninvited four-legged creatures from your garden. Before investing in fencing (the most effective, but also the most costly, option), it may be worth testing some of them to see how your own particular local pest population reacts. I've tried a great many supposedly pest-repellent materials through the years, and though they may be effective for a short period of time, all eventually failed under the pressure of hungry critters, even when we alternated their use. In the case of devices, the motion-activated sprinkler called the ScareCrow remains the single most effective deer-repellent device and allows us to garden fence-free (for now).

Most repellent materials work by offending deer with their strong smell. In this category are various bars of soap (Lifebuoy is the most often used, but Irish Spring and others are also said to work), deer predator urine (such as coyote), dried-blood products, Milorganite, and sprays of putrid eggs, garlic, and/or hot pepper, as well as sweaty T-shirts or hair from humans or pets. In many cases, the aroma fades or the product washes away in the rain and must be reapplied frequently. One very effective material, Tree Guard (a latex-based substance that lasts on plants for a long time and has an extremely bitter flavor), can't be used on edibles because the bitter substance enters the plant vascular system, resulting in bitter, inedible crops.

The device category includes tin pie plates that bang in the breeze, transistor radios, fake reptiles or owls, and motion-activated lights or sprinklers. In my experience, any of the above may work for short periods of time but could lose effectiveness over time as the critters adjust. If you choose to use devices or repellents rather than build a fence, take the time to learn about your local pest population and try various tactics, as needed. The risk is only the potential loss of crops.

Critter Trapping and Removal

Finally, it is important to address the touchy subject of trapping and removal, or eradication. There are many local ordinances and laws that may govern the trapping and relocation of wildlife. Shooting or otherwise killing as a solution is even more personal and local in nature. Using a live trap also creates the possibility of trapping something other than your intended target; if you do in fact succeed, you'll need to figure out what to do with your unexpected prisoner.

TOMATO DISEASES

The rapid popularity of tomatoes in America was soon followed by reports of disease. Alexander Livingston mentions a nonspecific "blight" beginning to cause trouble in his 1893 publication *Livingston and the Tomato*. Selective breeding for fusarium wilt resistance began in 1910 at the Tennessee and Louisiana agricultural stations. In the years since, scientific agencies have carried out work leading to the far more detailed and differentiated list of tomato diseases that we are aware of today. In general, tiny invading organisms that may be viral, bacterial, or fungal are the cause of tomato diseases. Some diseases are systemic and enter the plant from infected soil through the roots; others act on the plant surfaces and invade by soil splashed onto foliage or wind-blown infected soil. Still other diseases may get into the plant by the chewing action of insect carriers. In all cases, the impact on infected plants varies from reduced yield and tomato quality, accompanied by clearly troubled foliage and plant growth, to plant death, often before a single tomato is produced. The severity of the disease depends on the specific disease agent and timing of onset, combined with the weather conditions during infection.

As a rule, tomato diseases are stubborn and often fatal, and attempts at treatments for the vast majority have been met with little to no success. To make things worse, diseases tend to evolve over time, forming new strains, which render the plant breeder's attempts at packing resistance into a tomato's genes all the more difficult to sustain. Though chemical agents can delay the onset of diseases, I personally choose not to use them.

RESISTANCE AND TOLERANCE

Some tomato varieties have the ability to block infection by particular diseases; this is referred to as resistance (because the tomato resists the disease). Others become infected with the disease but still manage to perform well; this is referred to as tolerance. The presence of various letters next to tomato names in a seed catalog indicates resistance or tolerance to particular diseases. The letters most commonly used are F, which stands for fusarium, V for verticillium, and N for nematodes. (Nematodes, tiny worms that work at the plant roots to limit healthy growth, are not strictly a disease like fusarium and verticillium, but plants are able to have resistance or tolerance to their impact; see page 197 for more details.) Certain pathogens have evolved into different races specific to particular growing areas. Fusarium, for example, now exists in three races.

The presence of a letter next to the tomato name doesn't guarantee that the variety will thrive if grown in soil contaminated with the disease agent. What happens to the plants growing in the garden is determined by the three-way interaction of the pathogen, the plant, and the environmental conditions — this is known as the disease triangle. In certain conditions, the disease will overwhelm even the most resistant plants, but those with tolerance could survive sufficiently to obtain a decent crop.

For those gardeners who wish to focus on open-pollinated varieties (such as heirlooms), the lack of letters after the tomato names designating disease resistance doesn't mean that a variety has no tolerance or resistance to various diseases. It simply means that it has not been tested against them. A main reason that I grow so many different varieties is to hedge my bets. We love having lots of tomatoes each summer, and we love variety. If particular diseases affect a certain plant, it's likely that we'll still have plenty of tomatoes to eat from the more robust plants.

OUR DEER DILEMMA

Although we've battled deer for the 20 years we've lived in Raleigh, I can still sit on our deck and appreciate their beauty as they meander through the woods in back of our house. We still find it amusing when we walk our dogs and come upon a mini-herd of does with their fawns standing in a neighbor's front lawn, watching us — even when we're within just 10 feet of them.

As beautiful as they are, deer cause problems in the vegetable garden. Since deer are the tomato-damaging pest that I've had the most experience with through the years, I can share all of the many techniques that were ultimately unsuccessful, and the single technique that still, thankfully, seems to be working very well for me.

It's important to note that it took the deer a few years to find me. My first garden, 20 years ago, was in a side yard and unfenced. The deer actually gave us two or three years of unfettered tomato growing before we saw those first few telltale signs of nibbled plants. The situation grew rapidly worse: the deer inflicted significant damage on the plants and then the developing and ripe fruit started to vanish. Nothing causes more sadness in avid gardeners than seeing months of hard work disappear seemingly overnight, whether by disease or critter damage. And somehow, deer have a knack for knowing just which heirloom variety is held most dear, rather than targeting the common hybrid.

Sprinkling, spraying, hanging

We started dealing with the deer by degree and quickly found that many things will repel deer for short periods of time. Deer learn and adapt. Whatever method worked today will likely not work well in a few weeks. Here is what we tried before we fenced our garden: coyote urine, my urine, sweaty T-shirts, a transistor radio operating 24/7, Milorganite, dried blood, moth balls, dog hair, human hair, Irish Spring and Lifebuoy soap bars, and various things you hang on stakes or spray on the foliage made from hot pepper, garlic, or egg extracts. If anything, the coyote urine

and radio worked for the longest time, but even that was only a few weeks. And the most effective spray — Tree Guard — was also the worst solution. Apparently, the active compound, an impossibly bitter substance called Bitrex, gets into the vascular system of the plant. Young seedlings, pre-flower, sprayed with the Tree Guard withstood deer attacks, but the resulting tomatoes were too bitter to eat!

Physical barriers

Realizing that all repelling substances would have only limited success, we moved on to the fence solution. We started with monofilament fishing line at 2-, 4-, and 6-foot heights. This worked for almost an entire season, but eventually it failed, with the deer going under, through, or over the filament. The next fencing material, a 6-foot-tall plastic mesh, was effective once again for most of a season until the deer managed to jump it. We then moved to a battery-powered electric fence, with wires at 2, 4, and 6 feet above the ground. Perhaps it was the smell of ripening tomatoes that drove them to jump over the electrified 6-foot height — even after we hooked it up to a more powerful generator that was powered from house current. I think you get the idea: every attempt at fencing met with an adaptation of the deer to thwart the method after a period of time.

All of this messing around with fences coincided with a decline of our garden's soil. We were building up diseases in the soil and some of the surrounding trees were growing taller, so we decided to move our tomatoes to the sunniest area of our yard — the driveway. We also felt that the deer would be far less likely to maraud in an area much closer to the coming and goings of humans. How wrong we were! We ended up repeating much of the same dance we did with the first garden: no fence, repellent materials, monofilament fence, tall mesh fence, electric fence. The most remarkable moment came when I realized that the deer were actually bending down and crawling under the mesh and filament fence to get into the middle of the driveway garden.

Best deterrent . . . so far so good

I was at my wit's end by this time. The one solution I hadn't tried was using motion detector technology combined with a spray of water. A wonderful device called the ScareCrow seems to have solved our deer problem. A very sensitive motion detector with a surprising range triggers a quick blast of water. The reset time is quite rapid. The deer, for whatever reason, do not like a spontaneous bath, and for three years it has worked like a charm. We position two of the ScareCrows at the periphery of each of our gardens so that all approach lanes are covered. Except for the surprised (and wet) human visitors who have been blasted by our driveway devices, we are delighted with how these solved a persistent and depressing garden issue.

Since deer do have an amazing ability to adapt, I would not be surprised if the ScareCrow solution becomes less effective over time as more and more people resort to it. The best solution to deer will always be a good fence. Gardeners who can afford the cost and have no issues with neighborhood ordinances should invest in a fence of sturdy materials and a height of 8 feet; I've read numerous reports of deer easily jumping 6-foot fences. Such a fence could also help solve other four-legged critter issues, such as groundhogs and rabbits, as long as the fence runs down and into the soil for at least a foot. As the deer problem in many areas of the country will likely only increase with time, I'm sure more and more solutions (however temporary) will emerge.

The most important aspect of tomato disease identification is in knowing whether it will affect neighboring plants or carry through to saved seed. The main reasons for identifying the disease(s) affecting your tomato plants are control (for the current season) and planning for success the following season. If a particular disease easily spreads from plant to plant, early removal of infected plants could save those yet to be affected. Removal of lower, diseased foliage in some cases could help the plant survive and produce a reasonable yield. If it is clear that the disease entered the plant through infected soil, you will need to find a different tomato location the following year. Finally, if you need to save seed from the tomatoes of an infected plant, knowing if that particular disease will pass into saved seeds is very important.

Hygiene is also extremely important, especially to avoid damping-off during seed starting. Be sure to start with clean containers and sterile, lightly textured, quickly draining seed-starting medium. The sterility ensures that no fungal spores are present that can attack the young seedlings as they germinate.

Again, it's important to remember that disease identification is neither easy nor foolproof; a conversation with your local Extension agent is always the best course of action.

Natural Disease Control Options

COPPER — Copper fungicides can be effective in controlling some fungi and bacteria, but they can also injure the host plant if not used carefully. Two examples are copper sulfate and Bordeaux mixture (copper sulfate with lime). Copper compounds are diluted in water and sprayed on the plant as a preventive measure. (For all of these controls, it is important to carefully follow the directions on the products.)

SULFUR — This natural fungicide has been used for thousands of years. Sulfur can be applied as solutions or as dusts.

BICARBONATES — Baking soda is a well-known fungicide. It is dissolved in water and sprayed onto the plants.

NEEM OIL — Pressed from seeds and fruits of the neem tree, this oil is very useful on certain fungi and pests. It is typically found as a ready-to-spray solution, or a concentrate that you prepare yourself.

HORTICULTURAL OILS — Used as diluted sprays for the prevention of disease by controlling insects that spread them, horticultural oils are lightweight oils that may be either vegetable or petroleum based.

TEA TREE OIL — Also known as melaleuca oil, this essential oil is used in dilution with water for control of certain fungi, and also for whitefly.

ACTINOVATE — This recently available natural biological fungicide could be helpful with tomato fungal diseases as well as nematodes.

PROBLEM

Young leaves turn bronze, with small brown speckles. The plant becomes stunted and performs poorly. If fruit form, they exhibit round, pale concentric blemishes.

CULPRIT: Tomato spotted wilt virus, carried by thrips (members of the order Thysanoptera).

INSECT I.D.: Tiny, slender, flying insects with fringed wings. They're not particularly good flyers but can be carried by wind. Thrips chewing on tomato flowers transmit tomato spotted wilt virus (TSWV).

CONTROL/PREVENTION TACTICS

- Use a reflective mulch, such as aluminum foil.
- Position yellow sticky tape around the garden perimeter.
- Spray open blossoms with neem oil or insecticidal soap as a preventive measure.

PROBLEM

The foliage of young seedlings shows signs of being eaten.

VARIETIES OR PLANT PARTS MOST LIKELY AFFECTED: Young seedlings.

CULPRIT: Slugs, shell-less gastropod mollusks from a number of species.

INSECT I.D.: Soft-bodied slugs can change shape as they move, have prominent "antennae," and leave behind a slimy trail. Slugs cause significant destruction to plant parts or fruit of young plants. They are voracious nighttime feeders that must avoid the hot sun in order to survive. It's important to keep slugs away from young tomato plants, but they won't do significant damage to well-established plants.

CONTROL/PREVENTION TACTICS

- Examine trays of young seedlings for slugs hiding at the bottom.
- Ensure there is no smooth approach to the plant by using mulch or rough gravel.
- Set out a shallow saucer of beer for a trap.

PROBLEM

Foliage is speckled, discolored, and faded, and undersides show tiny mites and their webs.

CULPRIT: Spider mites, members of the Acari family.

INSECT I.D.: Tiny bodies of various colors found on the leaf undersides. They often spin webs.

CONTROL/PREVENTION TACTICS

- Introduce ladybugs as a natural predator.
- Spray affected areas with neem oil or insecticidal soap.
- In an extreme attack, remove and destroy affected foliage.

PROBLEM

After planting, the main stem of a young seedling is eaten at the soil line, toppling it.

CULPRIT: Cutworms, the larvae of various kinds of moths.

INSECT I.D.: Generally yellow, green, brown, or gray soft-bodied caterpillars, occasionally with longitudinal stripes, up to 1 inch long. Nocturnal feeders, cutworms eat through the stem at the soil line.

CONTROL/PREVENTION TACTICS

- To prevent damage, wrap the stem loosely in a barrier such as aluminum foil, with some both above and below the soil line, or fashion a cutworm collar out of cardboard encircling the just-planted seedling.

PROBLEM

Droopy, wilted-looking foliage.

VARIETIES OR PLANT PARTS MOST LIKELY AFFECTED: Paste or heart-shaped fruit, such as Anna Russian, Yellow Oxheart, and Opalka, are more likely to exhibit a genetic tendency to wilt.

POSSIBLE CULPRITS: Genetic tendency (in certain varieties), desiccation, poor drainage, chewing insects, walnut wilt.

CAUSES: Any plant (especially container-grown plants) will show some wilting during hot, dry days. Any plant that wilts even when well watered may suffer from poor drainage, which can lead to either rotting or disease of the roots, which can lead to plant death.

CONTROL/PREVENTION TACTICS

- Select non-wilting varieties.
- Plant in well-drained soil and supply adequate water.
- Examine young plants for aphids (see page 183 for control measures). Older plants may be afflicted with thrips (see page 179 for control measures), which spread tomato spotted wilt virus.
- Avoid planting near black walnut trees, which produce juglone, a substance that is toxic to tomato plants.

PROBLEM

Lower foliage turns yellow with age.

VARIETIES OR PLANT PARTS MOST LIKELY AFFECTED: Any plant.

CULPRITS: Early blight (alternaria fungus) or leaf spot (septoria fungus). Brown spotting earlier in the season is also associated with these.

CONTROL/PREVENTION TACTICS

- Remove and destroy lower yellow foliage.
- Mulch well to prevent soil from splashing onto foliage.

PROBLEM

General yellowing of foliage.

VARIETIES OR PLANT PARTS MOST LIKELY AFFECTED: Any plant.

CULPRIT: Could be associated with bacterial canker, bacterial pith necrosis, or verticillium or fusarium wilt.

CONTROL/PREVENTION TACTICS

- If the foliage is yellow without wilting and mainly on the lower parts of the plant, remove and destroy affected leaves.
- If the yellowing spreads rapidly upward and is accompanied by wilting, it is best to remove the plant.
- Mulching well to prevent soil from splashing on the foliage is recommended to minimize disease spread.
- Starting with disease-free seed, rotating beds away from areas with past issues, and bleaching/washing previously used containers, cages, and stakes is recommended.

PROBLEM SOLVING: FOLIAR DAMAGE

PROBLEM

Significant portions of leaves, stems, and fruit chewed away. Parts of the plant and fruit vanish overnight, with small fecal pellets found under damaged areas.

CULPRIT: Tomato hornworm, the larva of *Manduca quinquemaculata* (five-spotted hawk moth), or tobacco hornworm, the larva of *Manduca sexta* (Carolina sphinx moth).

INSECT I.D.: Large, thick green worms with eight pale V-shaped markings and black horns. The larvae can quickly reach 3 to 4 inches in length — not surprising, considering how much they eat. A relative, the tobacco hornworm, is similar but has seven diagonal lines and red horns.

CONTROL/PREVENTION TACTICS

- Locate, handpick, and destroy. (If a worm has the white rice-like egg cases of a parasitic wasp, leave it on the plant. The worm will soon die.)
- Spinosad is an effective organic control.

PROBLEM

Plant appears stunted, and foliage is distorted and mottled.

CULPRIT: Whiteflies, members of the family Aleyrodidae.

INSECT I.D.: Tiny white flying insects that reside on the underside of the foliage and suck on the plants' juices; they fly and create a small cloud when disturbed.

CONTROL/PREVENTION TACTICS

- Continually apply insecticidal soap to the affected areas.
- Place sticky yellow traps around the plant to trap them.
- Introduce natural predators such as ladybug and lacewing.
- Use a reflective mulch around the plant base, such as aluminum foil.
- In extreme cases, remove and destroy infected areas of the plant.

PROBLEM

Foliage has blotchy spots, and fruit may have small holes surrounded by a brown zone.

CULPRIT: Tomato pinworm, *Keiferia lycopersicella*, the larva of a moth.

INSECT I.D.: This pest spins a tent, and under this it tunnels into the foliage and sometimes the fruit. The tomato pinworm can be a problem in warm areas from Georgia southward, as far west as California, and occasionally in greenhouses farther north.

CONTROL/PREVENTION TACTICS

- Use organic controls, such as parasitic wasps and Bt spray.

PROBLEM

Foliage is mottling, curling, and wilting, along with reduced plant vigor and foliage that is covered with a sticky secretion or "honeydew."

CULPRIT: Aphids, members of the superfamily Aphidoidea.

INSECT I.D.: Small (1/8 inch long or less) sap-sucking insects of various colors.

CONTROL/PREVENTION TACTICS

- Introduce ladybug or lacewing larva.
- Spray with soapy water or insecticidal soap solutions.
- Apply pyrethrin dust.

PROBLEM

One stem/shoot of the plant turns bright yellow and wilts.

VARIETIES OR PLANT PARTS MOST LIKELY AFFECTED: Any plant.

CULPRIT: Fusarium wilt

CAUSE: A fungal infection brought on by wet soil and ambient temperatures between 80° and 90°F. The disease is soilborne and infects the plants through the roots. When the main stem holding the yellow foliage is cut, dark chocolate-brown streaks may be seen running lengthwise through the stem, often extending upward for some distance and especially evident at the point where the petiole joins the stem. Part of the plant may remain healthy and bear fruit, or the disease spreads and kills the entire plant.

CONTROL/PREVENTION TACTICS

- Start with seed that is certified to be free of the disease; don't save seeds from fusarium-infected plants.
- Avoid areas where fusarium has been a problem in the past.
- Bleach or otherwise sanitize containers, stakes, and cages prior to use.

PROBLEM SOLVING: FOLIAR DAMAGE

PROBLEM

Green and yellow variegated foliage.

VARIETIES OR PLANT PARTS MOST LIKELY AFFECTED: Any plant.

CULPRIT: Most often associated with viral issues, such as cucumber or tobacco mosaic virus.

CONTROL/PREVENTION TACTICS

- Viral issues are typically spread by chewing insects, such as aphids (see page 183) and striped cucumber beetles, so controlling these pests will limit the possibility of spreading the disease.
- Plant tolerant or resistant varieties.
- Remove affected plants.
- Rotate beds away from areas of infection.

PROBLEM

Purple foliage.

VARIETIES OR PLANT PARTS MOST LIKELY AFFECTED: Very young foliage; more mature plants may exhibit a purplish cast, particularly on the underside of foliage.

CULPRIT: Cool temperatures lead to a phosphorus deficit, which goes away as the weather warms and plants mature.

CONTROL/PREVENTION TACTICS

- No action necessary.

LESSER PESTS

The following are some common pests that frequently attack other plants in the garden but aren't likely to bother tomatoes.

JAPANESE BEETLES (*Popillia japonica*). These shiny brown beetles are commonly found in many gardens across North America.

FLEA BEETLES (members of the family Chrysomelidae). These tiny, shiny, nearly round brown beetles jump away quite quickly when disturbed — much like a flea, actually.

BLISTER BEETLES (beetles of the family Meloidae). These large and conspicuous insect pests are named for the skin-blistering agent cantharidin they emit when threatened.

COLORADO POTATO BEETLES (*Leptinotarsa decemlineata*). The sizable adult beetle has black and cream stripes and a deeper yellow head. It lays yellow to orange eggs on foliage undersides, which lead to hungry brownish larvae that consume the foliage.

LEAF MINERS (the larvae of various moths and flies). Small worms or insects that eat into foliage, leaving white tracks visible on the foliage.

CABBAGE LOOPER (*Trichoplusia ni*). The smooth, pale green, white-striped cabbage looper, also known as cabbage worm, is the larva of a moth.

PROBLEM

Brown or black foliage

VARIETIES OR PLANT PARTS MOST LIKELY AFFECTED: Any plant.

CULPRIT: Many diseases will present as a total browning or blackening of the foliage with age. Bacterial wilt and late blight tend to bypass foliage yellowing and move directly to browning.

CONTROL/PREVENTION TACTICS

- Mulch well to prevent soil from splashing on the foliage.
- Remove affected foliage.
- If the entire plant begins to wilt, remove the plant.

PROBLEM

Curled foliage.

VARIETIES OR PLANT PARTS MOST LIKELY AFFECTED: Most often found on the lower foliage of mature plants carrying a large fruit set, especially varieties that yield heavily, and indeterminate varieties as a class.

CULPRIT: Most often, this is a physiological issue that is generally not a cause for concern as the plant matures and conditions change. The problem tends to be worse when the plant is standing in poorly draining soil. Occurs most often in spring when there is a spell of wet weather.

CONTROL/PREVENTION TACTICS

- No action necessary, though take care to rule out leaf roll due to the presence of insects, drift of nearby herbicide onto the plant, or viral diseases, especially if the issue persists well into the season and plant health declines.

PROBLEM

Foliage becomes thin, twisted, and mottled with light and dark green streaks.

VARIETIES OR PLANT PARTS MOST LIKELY AFFECTED: Any plant.

CULPRIT: Tobacco mosaic virus (TMV).

CAUSE: Pathogen is spread by insects or seed, or it is skin-borne (from touching the plant) or carried in diseased roots from previous seasons.

CONTROL/PREVENTION TACTICS

- TMV can be seed-borne; don't save seeds from fruit from affected plants.
- Use resistant cultivars.
- TMV remains in plant debris for many years, so use good garden hygiene, and rotate beds away from problem areas.
- Remove and destroy infected plants.

PROBLEM

Leaves show yellowing with a V-shaped pattern, turning brown, with brown vascular tissue inside stem at crown of plant; plants wilt in the heat of the day.

VARIETIES OR PLANT PARTS MOST LIKELY AFFECTED: Any plant.

CULPRIT: Verticillium wilt.

CAUSE: A fungal infection brought on by wet soil; ambient temperature of 75°F is optimum for disease development. Fungus attacks the roots and works its way up the plant, which wilts later in the season. The fungus lives in the soil for a long time. To confirm infection, make a vertical slice of the main stem just above the soil line and look for brown parts in the conducting tissues under the skin, which will proceed upward as well as downward into the roots. Unlike fusarium wilt, verticillium wilt discoloration seldom extends more than 1 foot above the soil line.

CONTROL/PREVENTION TACTICS

- Grow tomatoes in well-drained soil.
- If verticillium wilt has been a problem in the past, try using varieties with tolerance or resistance, indicated by a "V" after the tomato name.
- Grow in areas away from the infected beds.
- Remove and destroy affected plants.

PROBLEM

Large, dark-colored, water-soaked lesions on foliage, with white spores developing at the edges. Fruit have large, firm, greasy, rough, brown spots with eventual white fungal spores.

VARIETIES OR PLANT PARTS MOST LIKELY AFFECTED: Any plant.

CULPRIT: Late blight.

CAUSE: Nights in the 50s, days in the 70s, with rain, fog, or dew. Infected living plants shed spores, which blow onto other plants and infect them. Late blight is one of the most devastating tomato diseases and it is well worth keeping up with the latest developments.

CONTROL/PREVENTION TACTICS

- Resistant cultivars are beginning to appear and current research will surely lead to more.
- If late blight is suspected, place infected plant parts in a ziplock bag and take them to a local Extension agent for confirmation.
- The infected plants and nearby plant material should be removed and destroyed. The disease does not overwinter in the soil.

PROBLEM

Foliage shows small dark spots, initially water soaked, coalescing and becoming angular, sometimes with a yellow halo; considerable loss of lower leaf foliage can occur. Fruit have small, brown, scabby, sunken or slightly raised spots.

VARIETIES OR PLANT PARTS MOST LIKELY AFFECTED: Any plant.

CULPRIT: Bacterial spot.

CAUSE: A bacterial infection brought on by wet conditions and temperatures between 75 and 85°F. The bacteria may be seed-borne. Entry of bacteria into plants occurs through natural plant openings or through wounds spread and caused by windblown soil, insects, or cultural practices.

CONTROL/PREVENTION TACTICS

- If this has been an issue in your garden, be sure to start with disease-free seed.
- Grow in areas away from any previously affected garden beds.
- Use good garden hygiene — clean up weeds and other garden plant debris.

PROBLEM

Mottled, ferny foliage that becomes extremely slender.

VARIETIES OR PLANT PARTS MOST LIKELY AFFECTED: Any plant.

CULPRIT: Cucumber mosaic virus.

CAUSE: Pathogen is spread by the chewing action of infected aphids, or by human contact if the virus is on the skin.

CONTROL/PREVENTION TACTICS

- Since the disease is spread by aphids, the only control is to minimize the presence of aphids on tomato plants (see page 183).
- Remove and destroy affected plants. The virus is not seed-borne and does not survive on plant debris.

PROBLEM

Leaves have tiny, dark brown to black spots with surrounding yellow halo; plant and foliage may be stunted, and fruit have tiny black spots that may be raised or embedded.

CULPRIT: Bacterial speck.

CAUSE: Wet conditions, between 65° and 75°F. Bacteria can be seed-borne or spread by infected transplants, as well as overwintered on infected weeds.

CONTROL/PREVENTION TACTICS

- If this has been an issue in your garden, be sure to start with disease-free seed.
- Grow in areas away from any previously affected garden beds.
- Use good garden hygiene — clean up weeds and other garden plant debris.

PROBLEM

The undersides of lower leaves develop small water-soaked spots with gray centers and brown edges. The spots enlarge and severely affected leaves yellow and drop.

VARIETIES OR PLANT PARTS MOST LIKELY AFFECTED: Any plant.

CULPRIT: A fungal infection called septoria leaf spot.

CAUSE: High humidity, temperatures between 65° and 75°F. Plants can be infected by soil splashing on lower foliage, infected equipment, hands, or chewing insects. The disease could possibly be seed-borne; not confirmed.

CONTROL/PREVENTION TACTICS

- Mulch to prevent soil from splashing onto lower foliage.
- Remove affected lower foliage periodically.

PROBLEM

Brown to black target-like spots on lower leaves, which eventually turn yellow and drop.

VARIETIES OR PLANT PARTS MOST LIKELY AFFECTED: Any plant.

CULPRIT: Early blight (a fungal infection caused by alternaria).

CAUSE: Wet conditions, temperatures between 75° and 85°F. Disease can overwinter on infected plant debris. May also be seed-borne.

CONTROL/PREVENTION TACTICS

- Mulch to prevent soil from splashing onto lower foliage.
- Remove affected lower foliage periodically.
- Start with disease-free seed.

PROBLEM

Developing fruit early in the season have blossom ends that are dark, almost waterlogged-looking, and quickly turn a greasy-leathery, dark blackish brown.

VARIETIES OR PLANT PARTS MOST LIKELY AFFECTED: Roma/paste varieties, and some of the longer indeterminate sauce types like Opalka and Speckled Roman. Adverse growing conditions can make it problematic for many other varieties, as well.

CULPRIT: Blossom end rot (BER).

CAUSES: Environmental conditions that affect the supply of water and calcium in the developing fruits, such as suddenly exposing rapidly growing plants to a period of drought; cultivation that damages roots; heavy, wet, cold soil; excess soluble salt in the soil.

CONTROL/PREVENTION TACTICS

- Test your soil to make sure the pH is adequate for tomatoes (6.2 to 6.8), and augment with lime (to add calcium and raise the pH) if necessary.
- Mulch around the base of the plants to conserve soil moisture during hot spells and water regularly, especially on very hot, dry days.
- Avoid varieties that seem to have consistent problems with BER.

MORE ABOUT BER

Often, BER will be limited to fruit on the first one or two blossom clusters, which are formed at a time when conditions promote BER formation. These fruits can simply be plucked and composted. BER can also occur at the center of the tomato. There are no indications externally for BER-afflicted large tomatoes that are spotless on the outside but rotten on the inside. It is in cutting a lovely, ripe, eagerly anticipated tomato that the dark, rotten tomato heart is then revealed. Tomatoes with the affliction are fine for seed saving and will not pass on the problem to subsequent generations.

PROBLEM

Poor fruit set.

VARIETIES OR PLANT PARTS MOST LIKELY AFFECTED: Large-fruited heirlooms.

CULPRIT: Poor fruit set is often very closely related to blossom drop.

CAUSES: Daytime temperatures above 90°F and nighttime temperatures above 70°F hinder pollination, leading to blossom drop rather than successful fruit set. Inadequate or excess nitrogen or very high or low humidity also contribute.

CONTROL/PREVENTION TACTICS

- Plant at the appropriate time; stagger seed starting to have plants at different stages through the season (to increase the odds of having plants blossom during periods of moderate weather).
- Prune fewer suckers to increase the odds of having enough blossoms set to achieve an adequate tomato crop.

PROBLEM

Holes bored into green tomatoes, which cause premature ripening and rapid rotting.

CULPRIT: Tomato fruitworm, the larva of *Helicoverpa zea,* a nondescript brownish moth. It also attacks cotton, where it takes the name cotton bollworm, and is found on corn, in which case it is referred to as the corn earworm.

INSECT I.D.: The worms may be green, brown, or pinkish with darker longitudinal stripes, and they grow to about 1½ inches long. Worms move quickly from fruit to fruit.

CONTROL/PREVENTION TACTICS

- Handpick and destroy, remove green fruit with holes from worms, use Bt spray or spinosad.

PROBLEM

A network of brown corky folds and creases at the blossom end of the tomato.

VARIETIES OR PLANT PARTS MOST LIKELY AFFECTED: Large-fruited beefsteak types that arise from large blossoms.

CULPRIT: A physiological condition called cat-facing.

CAUSE: Extended periods of relatively cool daytime (60° to 65°F) and nighttime (50° to 60°F) temperatures cause abnormal development of plant tissue between the style and ovary. Soil that is excessively high in nitrogen and overly aggressive pruning of the plant are additional causes.

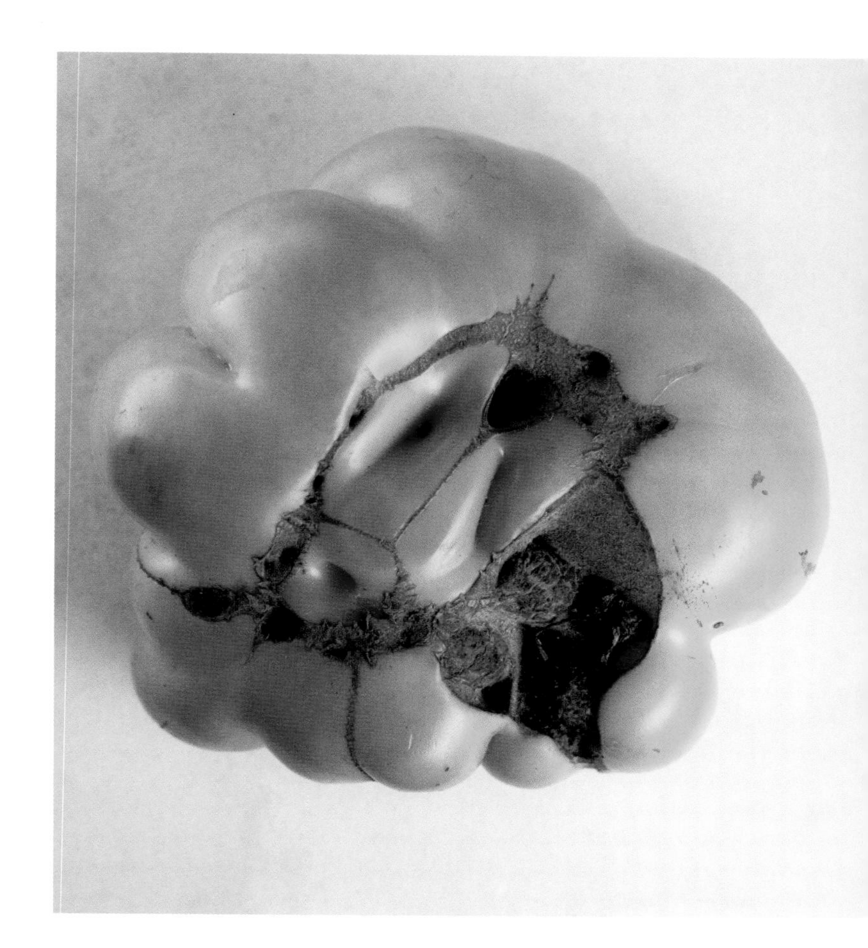

CONTROL/PREVENTION TACTICS

- Switch to smaller-fruited varieties, or try varieties that are described as being relatively smooth or round.
- Delay planting so that temperatures during flowering are warmer than the ranges that could lead to cat-facing.
- Avoid excess nitrogen and aggressive pruning.

BREEDING OUT CAT-FACING

During my hours spent reading old seed catalogs, I've found that cat-facing is one of the characteristics of many tomatoes that plant breeders were trying to eliminate from 1850 to 1880. It is, in a way, quite ironic that some of the most treasured tomatoes grown today are exactly like some of the most reviled (because of their irregularities, misshapenness, and ugly scars) tomatoes back when our grandparents, and their parents, were gardening. They wanted smooth, medium-size, canning-ready types; we now know some of the best-flavored varieties can be quite "ugly." Tastes certainly change with the times!

PROBLEM

Fruits with uneven color and bleached-out areas that are more prone to rotting.

VARIETIES OR PLANT PARTS MOST LIKELY AFFECTED: Severely pruned plants, or those that have lost foliage because of disease, are the most susceptible.

CULPRIT: Sunscald.

CAUSE: Scalding of the skin and wall of the tomato by the rays of the sun.

CONTROL/PREVENTION TACTICS

- In very sunny, hot climates, and if you have a history of experiencing sunscald, reduce the amount of pruning on the plants to provide some foliar shading for the tomatoes.

PROBLEM

Cracking.

VARIETIES OR PLANT PARTS MOST LIKELY AFFECTED: Concentric cracks (running in a circle, parallel, and around the stem attachment) are more common on smaller, rounder varieties, such as Marglobe. Radial cracks (those that radiate outward from and perpendicular to the tomato stem) seem to happen on large, often unusually colored heirlooms, such as Cherokee Purple. Thin-skinned cherry tomatoes such as Sweet Million and Sungold seem quite prone to cracking when fully ripe. Aside from the cosmetic issue, open cracks are areas for disease development, especially during times of high humidity.

CULPRIT: Genetics. Uneven moisture can also cause cracking, especially during times of rapid fruit growth. Frequently, the first few cracks seem to heal over and the tomato ripens normally.

CAUSE: Many tomato varieties simply have a genetic tendency to crack.

CONTROL/PREVENTION TACTICS

- Reduce watering as varieties approach ripeness.
- Pick tomatoes at half ripeness (especially if heavy rain is in the forecast) and allow them to fully ripen indoors.
- Pick nearly ripe fruit from plants before a heavy rain.
- Mulch the plants to maintain even soil moisture.
- If your favorite varieties have a tendency to crack, it may be something you just learn to live with.

PROBLEM

Poor pigment formation on fruit; pale, washed-out fruit color.

VARIETIES OR PLANT PARTS MOST LIKELY AFFECTED: Big Boy, Celebrity, Aker's West Virginia.

CAUSE: Daytime temperatures above 90°F. Lycopene (the main pigment in red varieties) formation in tomatoes is inhibited when temperatures reach this level.

CONTROL/PREVENTION TACTICS

- No action necessary, though if a hot spell is in the forecast, it may be wise to harvest partially ripe fruit and let them finish the process on the kitchen counter in order to achieve full color.

PROBLEM

Partially eaten ripe tomatoes, with bite marks.

VARIETIES OR PLANT PARTS MOST LIKELY AFFECTED: Any plant.

CULPRIT: Raccoons, opossums.

CONTROL/PREVENTION TACTICS

- Physical barrier or trap and remove (subject to local laws).

PROBLEM

Jagged, torn holes in ripe tomatoes.

VARIETIES OR PLANT PARTS MOST LIKELY AFFECTED: Any plant.

CULPRIT: Birds.

CONTROL/PREVENTION TACTICS

- When water is scarce, birds will resort to finding water where they can, and ripening tomatoes are likely an attractive target. Ensure there is a water source, such as a bird bath, nearby; if the problem is persistent, ensure there is a physical barrier around plants or fruit clusters.

PROBLEM

Ripe tomatoes are "stolen" or partially eaten; tomato foliage eaten.

VARIETIES OR PLANT PARTS MOST LIKELY AFFECTED: Any plant.

CULPRIT: Rabbits, groundhogs, or squirrels (especially during a drought).

CONTROL/PREVENTION TACTICS

- Physical barrier around the plants or fruit clusters, or erect a fence.
- Provide a nearby water source, especially in areas that are densely populated by squirrels and gardens that have been targeted in the past.

PROBLEM

Sudden disappearance of fruit and large parts of plants.

VARIETIES OR PLANT PARTS MOST LIKELY AFFECTED: Any plant.

CULPRIT: Deer (look for characteristic footprints and droppings).

CONTROL/PREVENTION TACTICS

- The numerous deterrents are highly variable from season to season and from area to area, and they depend on local populations and weather. The only sure thing is a fence of appropriate height, or an electrified version. Refer to page 176 for more on my own experience with deer and success in deterring them with a motion-activated sprinkler.

PROBLEM

Ripe fruit develop sunken brown spots.

VARIETIES OR PLANT PARTS MOST LIKELY AFFECTED: Any plant.

CULPRIT: Anthracnose.

CAUSE: The spores are splashed onto fruit from infected soil. Infection is more likely when temperatures are above 80°F. The disease is more prevalent when the tomatoes are grown in poorly draining soil.

CONTROL/PREVENTION TACTICS

- Plant tomatoes in areas that drain well.
- Mulch well to prevent soil from splashing onto lower foliage and fruit.
- Rotate beds to a different area if this is a significant issue.

PROBLEM

Upon ripening, tomatoes show small, white, hard spots on the surface and into the flesh.

CULPRIT: Brown marmorated stinkbugs (*Halyomorpha halys*), an invasive species originally from China that was accidentally introduced into the United States in 1998.

INSECT I.D.: Brown, shield-shaped beetles, ½ to ¾ inch long and nearly the same in width, often with subtle markings, alternating light bands on the antennae, and alternating darker markings on the abdomen edge. Beetles suck on the fruit.

CONTROL/PREVENTION TACTICS

- Handpick and carefully destroy (when they're crushed, the odor is quite offensive).

PROBLEM

Newly germinated seedlings pinch off at the soil line, topple over, and shrivel up.

VARIETIES OR PLANT PARTS MOST LIKELY AFFECTED: Seedlings, just after germination.

CULPRIT: Damping-off fungus.

CAUSE: Contributing factors include water-logged soil or poor air circulation in the presence of the fungal spores from *Pythium* species or *Rhizoctonia solani.*

CONTROL/PREVENTION TACTICS

- Prevent infection by thoroughly sanitizing all equipment before starting seeds.

PROBLEM

Twisted or distorted growth.

VARIETIES OR PLANT PARTS MOST LIKELY AFFECTED: New growth and young transplants are most affected.

CULPRIT: Droplets of herbicide blow onto your plants on a breezy day, perhaps from a neighboring yard or field. Often many plants in a particular area of the garden will exhibit distorted growth, ruling out diseases such as cucumber and tobacco mosaic virus, which present in a similar way but are more random (and more rare).

CONTROL/PREVENTION TACTICS

- If you notice herbicide drift, a good immediate step is to hose down your plants to remove the chemicals from the leaf surface.

PROBLEM

Plants vanish into holes that are connected to "tunnels."

CULPRIT: Voles.

CONTROL/PREVENTION TACTICS

- Prevent future damage by digging a sharp, rocky material such as Permatill into soil around the plant root area. This will provide a physical challenge to the voles and they will tunnel elsewhere (voles simply will not dig through coarse-textured materials). It also provides nice aeration to the soil as a bonus.

PROBLEM

Raised areas in the yard or garden, and plants die from being displaced or toppled.

VARIETIES OR PLANT PARTS MOST LIKELY AFFECTED: Any plant.

CULPRIT: Moles.

CONTROL/PREVENTION TACTICS

- Spread castor oil–infused pellets around the garden as a repellent.

MOLE OR VOLE?

Voles look essentially like brownish mice, enjoy feasting on plants, and are more difficult to control than moles are. Moles look like little fuzzy gray sausages with short legs and a stubby tail. They hunt for grubs, and although they aren't interested in your tomatoes, they can disturb plants by dislodging root systems.

PROBLEM

Plants wilt and die quickly without foliage yellowing or spotting.

VARIETIES OR PLANT PARTS MOST LIKELY AFFECTED: Any plant.

CULPRIT: Bacterial wilt.

CAUSE: Wet conditions, with temperatures above 75°F. Bacteria enter roots in infected soils, spread from infected transplants, and can be spread by infected chewing insects.

To diagnose accurately, peel back and examine part of the stem above the soil line (early on the center is water soaked; later on it is brown or even hollow) or cut off a stem and place in a glass of water (the bacteria will ooze out as cloudy material).

CONTROL/PREVENTION TACTICS

- Remove affected plants immediately and destroy.
- Grow in areas away from beds that were previously affected.
- Grow in containers with fresh potting mix.

PROBLEM

Plants appear stunted or wilted, with swelling (or galls) on the roots.

CULPRIT: Root knot nematodes, tiny parasitic worms from the genus *Meloidogyne* that live in the soil in areas with warm or short winters.

INSECT I.D.: The worms are microscopic; presence of root galls is the best visual evidence.

CONTROL/PREVENTION TACTICS

- Grow varieties with resistance or tolerance (look for the letter "N" after the variety name).
- Note garden areas where the issues occur and avoid (only a temporary solution).
- Solarize soil, or plant the bed with marigolds for a season.

250 RECOMMENDED Tomatoes

WITH THOUSANDS OF TOMATO varieties to choose from, where does one start? The following list contains some of my favorites, organized by color, as well as some recent developments, varieties that are historically important, and those that are very popular and widely adapted. Consider it a starting point for your own tomato explorations.

TOMATO VARIETIES: RED

NAME	SHAPE	SIZE	SEASON	GROWTH	F_1 OR OP	FLAVOR
Abraham Lincoln	oblate	large	late	ind	OP*	balanced
Ace	oblate	medium	late	ind	OP	balanced
Aker's West Virginia	oblate	large	mid	ind	OP*	balanced
Amish Paste	paste	medium	mid	ind	OP*	balanced
Andrew Rahart's Jumbo Red	oblate	large	late	ind	OP*	balanced
Beefsteak	oblate	large	late	ind	OP*	mild
Better Boy	oblate	large	mid	ind	F_1	balanced
Better Bush	oblate	medium	mid	dwarf	F_1	balanced
BHN 589	globe	medium	mid	det	F_1	mild
Big Beef	oblate	large	mid	ind	F_1	balanced
Big Boy	globe	large	late	ind	F_1	balanced
Bisignano #2	paste	medium	mid	ind	OP*	intense
Bonny Best	globe	medium	mid	ind	OP*	tart
Box Car Willie	globe	medium	mid	ind	OP*	balanced
Break O'Day	globe	medium	late	ind	OP*	tart
Bulgarian Triumph	globe	medium	mid	ind	OP*	balanced
Celebrity	oblate	medium	mid	semi	F_1	balanced
Chalk's Early Jewel	globe	medium	mid	ind	OP*	tart
Colossal Red, improved	oblate	large	late	ind	OP*	balanced
Costoluto Genovese	oblate	medium	late	ind	OP*	balanced

KEY

F_1 — seed is an F_1 hybrid

OP — open pollinated, a.k.a. non-hybrid; those with a * are considered heirlooms

semi — semideterminate

ind — indeterminate

det — determinate

The colors indicated here refer to *apparent* color, determined by the combination of skin over flesh colors, except in the case of green tomatoes (all have green flesh; those with a + have amber skin when ripe).

Season: because of the wide variations in typical days-to-maturity information, "early" equates roughly to 45–60 days from transplant, "midseason" 60–75 days, and "late" 75 days or more.

Big Boy

TOMATO VARIETIES: RED

NAME	SHAPE	SIZE	SEASON	GROWTH	F_1 OR OP	FLAVOR
Creole	globe	medium	mid	ind	OP*	balanced
Cuostralee	oblate	large	late	ind	OP*	balanced
Delicious	oblate	large	late	ind	OP	mild
Druzba	oblate	medium	mid	ind	OP*	balanced
Dwarf Stone	oblate	medium	late	dwarf	OP*	balanced
Earliana	oblate	medium	early	ind	OP*	tart
Early Girl	oblate	medium	early	ind	F_1	balanced
Favorite	oblate	medium	mid	ind	OP*	balanced
Gallo Plum	pepper	medium	mid	ind	OP*	balanced
German Red Strawberry	heart	large	late	ind	OP*	balanced
Giant Syrian	heart	large	mid	ind	OP*	balanced
Grape	grape	small	early	ind	F_1	sweet

Giant Syrian

Favorite

NAME	SHAPE	SIZE	SEASON	GROWTH	F_1 OR OP	FLAVOR
Gurney Girl	globe	medium	mid	ind	F_1	balanced
Homestead	globe	medium	mid	ind	OP	balanced
Iditarod Red	globe	medium	early	dwarf	OP	mild
Jersey Devil	pepper	medium	mid	ind	OP*	balanced
Jet Star	globe	medium	mid	ind	F_1	balanced
John Baer	globe	medium	mid	ind	OP*	tart
Juliet	grape	small	early	ind	F_1	sweet
Kimberly	globe	small	early	ind	OP*	balanced
Lillian's Red Kansas Paste	paste	medium	mid	ind	OP*	intense
Marglobe	globe	medium	mid	ind	OP*	balanced
Marmande	oblate	medium	late	semi	OP*	balanced
Martino's Roma	pear	medium	mid	det	OP*	mild

Lillian's Red

TOMATO VARIETIES: RED

NAME	SHAPE	SIZE	SEASON	GROWTH	F_1 OR OP	FLAVOR
Matchless	oblate	medium	mid	ind	OP*	balanced
Matt's Wild Cherry	cherry	tiny	early	ind	OP*	sweet
Mayo's Delight	heart	large	late	ind	OP*	balanced
Mexico Midget	cherry	tiny	early	ind	OP*	intense
Moreton	globe	medium	mid	ind	F_1	balanced
Mule Team	globe	medium	mid	ind	OP*	balanced
Nepal	globe	medium	late	ind	OP*	intense
Neves Azorean Red	oblate	large	late	ind	OP*	balanced
Opalka	pepper	medium	late	ind	OP*	balanced
Oregon Spring	oblate	medium	early	det	OP	mild
OTV Brandywine	oblate	large	late	ind	OP	intense
Peron	globe	medium	late	ind	OP*	balanced
Prue	pepper	medium	mid	ind	OP*	balanced
Red Brandywine	oblate	medium	mid	ind	OP*	balanced
Red Pear	pear	small	early	ind	OP*	tart
Red Robin	cherry	small	early	dwarf	OP	mild
Reif Italian Red Heart	heart	medium	late	ind	OP*	balanced
Riesentraube	cherry	small	mid	ind	OP*	sweet
Roma	pear	medium	mid	det	OP	mild
Russian 117	heart	large	late	ind	OP*	intense
Rutgers	oblate	medium	mid	ind	OP*	balanced

Red Brandywine

Mexico Midget

TOMATO VARIETIES: RED

NAME	SHAPE	SIZE	SEASON	GROWTH	F_1 OR OP	FLAVOR
San Marzano	paste	medium	late	ind	OP*	mild
Scarlet Topper	oblate	medium	mid	det	OP*	tart
Shannon's	paste	medium	mid	ind	OP*	balanced
Silvery Fir Tree	oblate	medium	early	det	OP*	tart
Sophie's Choice	oblate	medium	mid	det	OP*	mild
Stick	globe	small	mid	ind	OP	mild
Stone	globe	medium	mid	ind	OP*	tart
Stupice	globe	small	early	ind	OP*	balanced
Success	globe	medium	mid	ind	OP*	balanced
Supersonic	oblate	large	late	ind	F_1	balanced
Sweet 100	cherry	small	early	ind	F_1	sweet
Sweet Million	cherry	small	early	ind	F_1	sweet
Tommy Toe	cherry	small	early	ind	OP*	sweet
Ultra Boy	oblate	large	late	ind	F_1	balanced
Variegated	globe	medium	early	ind	OP	mild
Wayahead	oblate	medium	early	det	OP*	tart
Wes	heart	large	late	ind	OP*	balanced
Whopper	oblate	large	late	ind	F_1	balanced
Zogola	oblate	large	late	ind	OP*	intense

TOMATO VARIETIES: PINK

NAME	SHAPE	SIZE	SEASON	GROWTH	F_1 OR OP	FLAVOR
Anna Russian	heart	medium	early	ind	OP*	sweet
Arkansas Traveler	globe	medium	mid	ind	OP	sweet
Aunt Ginny's Purple	oblate	large	mid	ind	OP*	sweet
Beauty	oblate	medium	mid	ind	OP*	balanced
Brimmer	oblate	large	late	ind	OP*	sweet
Burgundy Traveler	oblate	medium	mid	ind	OP*	balanced
Caspian Pink	oblate	large	mid	ind	OP*	sweet
Crnkovic Yugoslavian	oblate	large	late	ind	OP*	sweet
Dester	oblate	large	mid	ind	OP*	intense
Dr. Carolyn Pink	cherry	small	early	ind	OP	sweet
Dwarf Arctic Rose	oblate	medium	early	dwarf	OP	balanced
Dwarf Champion	oblate	medium	late	dwarf	OP*	sweet
Eva Purple Ball	globe	medium	mid	ind	OP*	sweet
Ferris Wheel	oblate	large	late	ind	OP*	intense
German Head	oblate	large	late	ind	OP*	balanced

Anna Russian

Ferris Wheel

Peak of Perfection

Polish

TOMATO VARIETIES: PINK

NAME	SHAPE	SIZE	SEASON	GROWTH	F_1 OR OP	FLAVOR
German Johnson	oblate	large	mid	ind	OP*	mild
German Pink	oblate	large	late	ind	OP*	sweet
German Queen	oblate	large	late	ind	OP*	sweet
Gregori's Altai	oblate	medium	early	ind	OP*	sweet
June Pink	oblate	medium	early	ind	OP*	balanced
Kalman's Hungarian Pink	heart	large	mid	ind	OP*	sweet
Kosovo	heart	large	mid	ind	OP*	sweet
Large Pink Bulgarian	oblate	large	mid	ind	OP*	sweet
Magnus	globe	medium	late	ind	OP*	balanced
Marianna's Peace	oblate	large	late	ind	OP*	sweet
Mortgage Lifter	oblate	large	late	ind	OP*	sweet
New Big Dwarf	oblate	medium	mid	dwarf	OP*	intense
Nicky Crain	heart	large	late	ind	OP*	mild
Olena Ukrainian	oblate	large	mid	ind	OP*	sweet
Omar's Lebanese	oblate	large	late	ind	OP*	sweet
Oxheart	heart	large	late	ind	OP*	mild
Peach Blow Sutton	globe	medium	mid	ind	OP*	sweet
Peak of Perfection	oblate	large	late	ind	OP*	intense
Pink Brandywine	oblate	large	late	ind	OP*	intense
Pink Ping Pong	globe	medium	early	ind	OP*	sweet
Pink Sweet	oblate	large	mid	ind	OP*	sweet
Polish	oblate	large	mid	ind	OP*	intense

TOMATO VARIETIES: PINK

NAME	SHAPE	SIZE	SEASON	GROWTH	F_1 OR OP	FLAVOR
Ponderosa	oblate	large	late	ind	OP*	sweet
Porter	oval	small	mid	ind	OP*	sweet
Pruden's Purple	oblate	large	mid	ind	OP*	balanced
Sandul Moldovan	oblate	large	mid	ind	OP*	intense
Soldacki	oblate	large	mid	ind	OP*	balanced
Stump of the World	oblate	large	late	ind	OP*	intense
Sweet Quartz	cherry	small	early	ind	F_1	sweet
Tappy's Finest	oblate	large	mid	ind	OP*	balanced
Tiffen Mennonite	oblate	large	late	ind	OP*	balanced
Ukrainian Heart	heart	large	late	ind	OP*	balanced
Watermelon Beefsteak	oblate	large	late	ind	OP*	balanced
Winsall	oblate	large	late	ind	OP*	balanced
Yukon Quest	oblate	medium	early	dwarf	OP	balanced
Zapotec Pink Ribbed	oblate	medium	mid	ind	OP*	mild

Stump of the World

TOMATO VARIETIES: PURPLE

NAME	SHAPE	SIZE	SEASON	GROWTH	F_1 OR OP	FLAVOR
Black Cherry	cherry	small	early	ind	OP	intense
Black from Tula	oblate	large	early	ind	OP*	balanced
Black Krim	oblate	large	mid	ind	OP*	tart
Brad's Black Heart	heart	large	late	ind	OP	balanced
Carbon	globe	large	mid	ind	OP	balanced
Cherokee Purple	oblate	large	mid	ind	OP*	intense
Dwarf Wild Fred	oblate	large	mid	dwarf	OP	balanced
Gary O'Sena	oblate	large	late	ind	OP	balanced
Indian Stripe	oblate	large	late	ind	OP*	intense
JD Special C Tex	oblate	large	late	ind	OP	intense
Perth Pride	globe	medium	mid	dwarf	OP*	intense
Price's Purple	oblate	large	late	ind	OP*	intense
Purple Calabash	oblate	large	late	ind	OP*	intense
Purple Russian	pepper	medium	mid	ind	OP*	balanced
Rosella Purple	oblate	large	mid	dwarf	OP	intense
Southern Night	oblate	medium	mid	det	OP*	balanced

Cherokee Purple

Rosella Purple

TOMATO VARIETIES: BROWN

NAME	SHAPE	SIZE	SEASON	GROWTH	F_1 OR OP	FLAVOR
Amazon Chocolate	oblate	large	late	ind	OP	balanced
Black Plum	pear	small	early	ind	OP*	tart
Black Prince	globe	medium	early	ind	OP*	sweet
Cherokee Chocolate	oblate	large	mid	ind	OP	intense
Japanese Trifele Black	pear	medium	mid	ind	OP*	balanced
Paul Robeson	oblate	large	early	ind	OP*	tart
Sleeping Lady	oblate	medium	early	dwarf	OP	balanced
Tasmanian Chocolate	oblate	large	mid	dwarf	OP	balanced

Cherokee Chocolate

TOMATO VARIETIES: YELLOW

NAME	SHAPE	SIZE	SEASON	GROWTH	F_1 OR OP	FLAVOR
Anna Banana Russian	heart	medium	mid	ind	OP	sweet
Azoychka	oblate	medium	early	ind	OP*	tart
Dwarf Sweet Sue	globe	medium	late	dwarf	OP	intense
Galina	cherry	small	early	ind	OP*	sweet
Garden Peach	globe	small	early	ind	OP*	mild
Gold Ball	globe	small	early	ind	OP*	mild
Gold Nugget	cherry	small	early	det	OP	mild
Golden Dwarf Champion	oblate	medium	mid	dwarf	OP*	sweet
Golden Queen	oblate	medium	mid	ind	OP*	balanced
Hugh's	oblate	large	late	ind	OP*	balanced
Isbell's Golden Colossal	oblate	large	late	ind	OP*	balanced
Lemmony	oblate	medium	mid	ind	OP*	balanced
Lemon Boy	oblate	medium	mid	ind	F_1	balanced
Lillian's Yellow Heirloom	oblate	large	late	ind	OP*	intense

Dwarf Sweet Sue

Golden Queen

TOMATO VARIETIES: YELLOW

NAME	SHAPE	SIZE	SEASON	GROWTH	F_1 OR OP	FLAVOR
Madara	cherry	small	early	ind	OP*	sweet
Manyel	globe	medium	mid	ind	OP*	balanced
Plum Lemon	paste	medium	mid	ind	OP*	mild
Summer Sunrise	oblate	medium	late	dwarf	OP	intense
Summertime Gold	oblate	medium	late	dwarf	OP	intense
Taxi	globe	medium	early	det	OP	mild
Yellow Bell	pear	small	mid	ind	OP*	balanced
Yellow Pear	pear	small	early	ind	OP*	mild
Yellow Stuffer	oblate	medium	mid	ind	OP*	mild

Hugh's

Lillian's Yellow Heirloom

TOMATO VARIETIES: ORANGE

NAME	SHAPE	SIZE	SEASON	GROWTH	F_1 OR OP	FLAVOR
Amana Orange	oblate	large	mid	ind	OP*	mild
Aunt Gertie's Gold	oblate	large	late	ind	OP*	intense
Dr. Wyche's Yellow	oblate	large	mid	ind	OP*	balanced
Earl of Edgecomb	oblate	medium	mid	ind	OP*	tart
Golden Gem	cherry	small	early	ind	F_1	intense
Jaune Flamme	globe	small	mid	ind	OP*	tart
Jubilee	oblate	medium	mid	ind	OP*	balanced
KBX	oblate	large	mid	ind	OP	intense
Kellogg's Breakfast	oblate	large	late	ind	OP*	balanced
Nebraska Wedding	oblate	large	mid	ind	OP*	mild

Golden Gem

Jubilee

TOMATO VARIETIES: ORANGE

NAME	SHAPE	SIZE	SEASON	GROWTH	F_1 OR OP	FLAVOR
Orange Banana	paste	medium	mid	ind	OP	balanced
Orange Heirloom	oblate	large	late	ind	OP*	balanced
Orange Minsk	oblate	medium	mid	ind	OP*	balanced
Orange Strawberry	heart	large	late	ind	OP*	balanced
Persimmon	oblate	large	late	ind	OP*	mild
Sun Gold	cherry	small	early	ind	F_1	intense
Sunray	oblate	medium	mid	ind	OP	balanced
Sweet Gold	cherry	small	early	ind	F_1	intense
Valencia	oblate	medium	mid	ind	OP	balanced
Yellow Brandywine	oblate	large	late	ind	OP*	tart
Yellow Oxheart	heart	large	late	ind	OP*	balanced

Yellow Oxheart

TOMATO VARIETIES: WHITE

NAME	SHAPE	SIZE	SEASON	GROWTH	F_1 OR OP	FLAVOR
Coyote	cherry	tiny	early	ind	OP*	sweet
Cream Sausage	paste	medium	mid	ind	OP	mild
Dr. Carolyn	cherry	small	early	ind	OP	sweet
Dwarf Mr. Snow	oblate	medium	late	dwarf	OP	intense
Great White	oblate	large	late	ind	OP	balanced
Lemon Drop	cherry	small	early	ind	OP	tart
Potato Leaf White	oblate	medium	mid	ind	OP*	mild
White Beauty	oblate	medium	late	ind	OP*	mild
White Oxheart	heart	medium	mid	ind	OP	mild
White Queen	oblate	medium	mid	ind	OP*	mild
Yellow White	oblate	medium/ large	late	ind	OP*	mild

Dwarf Mr. Snow

TOMATO VARIETIES: GREEN

NAME	SHAPE	SIZE	SEASON	GROWTH	F_1 OR OP	FLAVOR
Aunt Ruby's German Green	oblate	large	late	ind	OP*	intense
Cherokee Green+	oblate	large	mid	ind	OP	intense
Dorothy's Green+	oblate	medium	mid	ind	OP*	intense
Dwarf Beryl Beauty	globe	medium	mid	dwarf	OP	sweet
Dwarf Emerald Giant	oblate	large	late	dwarf	OP	balanced
Dwarf Jade Beauty	globe	medium	mid	dwarf	OP	sweet
Dwarf Kelly Green	globe	medium	mid	dwarf	OP	sweet
Evergreen+	oblate	medium	mid	ind	OP*	intense
Green Doctor	cherry	small	early	ind	OP	tart
Green Giant	oblate	large	mid	ind	OP	intense
Green Grape	cherry	small	early	det	OP	tart
Lime Green Salad	oblate	medium	early	dwarf	OP	tart
Summertime Green	oblate	large	late	dwarf	OP	intense

+ — These varieties have amber skin when ripe.

Green Giant

TOMATO VARIETIES: MULTICOLOR

NAME	SHAPE	SIZE	SEASON	GROWTH	F_1 OR OP	FLAVOR
Ananas Noire	oblate	large	late	ind	OP	balanced
Big Rainbow	oblate	large	late	ind	OP*	mild
Georgia Streak	oblate	large	late	ind	OP*	mild
Hillbilly	oblate	large	late	ind	OP	mild
Isis Candy	cherry	small	mid	ind	OP	sweet
Little Lucky	globe	medium	mid	ind	OP	intense
Lucky Cross	oblate	large	late	ind	OP	intense
Marizol Gold	oblate	large	late	ind	OP*	mild
Old German	oblate	large	late	ind	OP*	mild
Orange Russian 117	heart	large	late	ind	OP	mild
Pineapple	oblate	large	late	ind	OP*	mild
Regina's Yellow	oblate	large	late	ind	OP*	mild
Ruby Gold	oblate	large	late	ind	OP*	mild
Virginia Sweets	oblate	large	late	ind	OP*	mild

Ruby Gold

TOMATO VARIETIES: STRIPED

NAME	SHAPE	SIZE	SEASON	GROWTH	F_1 OR OP	FLAVOR
Berkeley Tie Dye	oblate	large	mid	ind	OP	balanced
Black Zebra	globe	medium	mid	ind	OP	tart
Blush	oval	small	early	ind	OP	sweet
Copia	oblate	large	mid	ind	OP	mild
Green Zebra	globe	medium	mid	ind	OP	tart
Maglia Rose	oval	small	early	ind	OP	sweet
Pink Berkeley Tie Dye	oblate	large	mid	ind	OP	intense
Red Zebra	globe	medium	mid	ind	OP	tart
Speckled Roman	pepper	medium	mid	ind	OP	balanced
Striped Cavern	globe	medium	mid	ind	OP*	mild
Striped German	oblate	large	late	ind	OP*	mild
Tiger Tom	globe	small	early	ind	OP*	tart
Tigerella	globe	small	early	ind	OP*	tart
Vintage Wine	oblate	large	mid	ind	OP	mild

Red Zebra

Tiger Tom

Oxheart

Golden Queen

SIGNIFICANT LIVINGSTON INTRODUCTIONS

PARAGON (1870) — This medium-size red variety originated from a single plant growing in a field of various varieties of that time.

ACME (1875) — Like Paragon, this medium-size pink tomato resulted from a selection from a single plant growing in one of Livingston's fields of various tomatoes.

PERFECTION (1880) — A medium-size, red-fruited tomato selected from Livingston's fields of Acme.

GOLDEN QUEEN (1882) — Originated from a yellow tomato Livingston acquired at a country fair.

FAVORITE (1883) — A medium-size, red variety that originated from Livingston's Paragon fields.

BEAUTY (1885) — A medium-size, pink-fruited selection from Livingston's Paragon fields.

POTATO LEAF (1887) — Livingston notes that he had this tomato ready for release prior to Beauty, and it is assumed to be a selection he made from fields of an unspecified variety.

STONE (1889) — A medium-size, red-fruited variety that was growing between rows of Beauty and Favorite.

ROYAL RED (1891) — A medium-size, red-fruited variety discovered in a planting of Dwarf Champion. Since Royal Red seems to be extinct, it is unclear as to whether it shared the dwarf growth habit.

GOLD BALL (1892) — Supposedly found in a grower's field of unspecified varieties, Gold Ball is a small, round, bright yellow tomato.

BUCKEYE STATE (1893) — Livingston claims he had this selected from the very beginning at the time he was completing his work on Paragon prior to 1870, so we assume it was a distinct plant growing in his fields of mixed varieties.

DWARF ARISTOCRAT (1893) — Livingston does not detail the history of this dwarf-growing, medium-size, red-fruited variety, but it is likely to have been a selection out of the pink-fruited Dwarf Champion fields.

HONOR BRIGHT (1898) — Described as a very unusual mutation, this variety has foliage that fades to yellow (despite being healthy), with unripe fruit that turn pale green to white, then to yellow and orange, and finally to red.

LARGE ROSE PEACH (1898) — Peach tomatoes were long known, but Livingston's selection was described as significantly larger; no other information of origin is available, so we can assume it was a selection from a field of the existing peach-type tomatoes of the time.

MAGNUS (1900) — Its origin is unknown, but it is likely a reselection of the older Livingston variety Potato Leaf released under a different name. The tomatoes are medium-size, round, and pink, and they are borne on a potato-leaf plant.

DWARF STONE (1902) — Unlike the other dwarf-growing tomatoes of its time, Dwarf Stone's red fruit are significantly larger, approaching 8 ounces.

GLOBE (1905) — This tomato, which became a very important shipping variety, originated with a cross between Stone and Ponderosa made in 1899. The medium-size pink fruits are almost perfectly round.

OXHEART (1925) — This tomato was likely a shape mutation found by a customer. It was the first example of a large, heart-shaped tomato.

APPENDIX REGIONAL GROWING TIPS

Alaska

GARDENERS: Susan Bailey and Sherry Shiesl

LOCATION: Anchorage and Wasilla, Alaska

GROWING SEASON: 100 days

AVERAGE SEASONAL HIGH TEMPERATURES: Low 70s

PEST/DISEASE PRESSURE: Almost none. Susan and Sherry's plants sometimes experience botrytis mold during wet weather in August.

SPECIAL CHALLENGES: A very short growing season.

TIPS: Start seeds in February. Grow in containers, on an asphalt driveway for additional heat, if possible. Get a greenhouse, if you can. Grow varieties that mature early and can tolerate cooler temperatures.

BEST VARIETIES: Because of the efforts of Sherry and Susan, gardeners can grow the wonderful new dwarf varieties Iditarod Red, Yukon Quest, Sleeping Lady, and Dwarf Arctic Rose, which are particularly good for short-season areas. Among varieties Sherry enjoys growing are Cherokee Purple, Native Sun, Polar Baby, Stupice, and Sophie's Choice. Susan's regulars include a number of Russian varieties that thrive in a similar climate, as well as Vorlon, Native Sun, Amber, and Aurora.

Pacific Northwest

GARDENERS: Tatiana Kouchnareva (owner of Tatiana's TOMATObase; see page 235) and Denise Salmon

LOCATION: Anmore, BC, and Vancouver, BC

GROWING SEASON: July and August

AVERAGE SEASONAL HIGH TEMPERATURE: 73°F

PEST/DISEASE PRESSURE: Slugs, thrips, and rodents, such as rats and squirrels, are the major pests. Gray mold (early spring, then at the end of the season) and powdery mildew, along with devastating, seemingly annual attacks of late blight, also cause problems. Foliage issues are exacerbated by the heavy dew that is experienced for much of the summer. Excess rain also leads to splitting fruit. Tatiana notes that it helps to grow in pots on the deck or balcony under a roof overhang to protect from rain and provide additional warmth from a house wall. It's also important in this climate to provide adequate ventilation between plants. As in any area that experiences diseases such as late blight, good gardening hygiene — including cleaning up garden debris at the end of the season — is critical for the success of the next year's efforts.

SPECIAL CHALLENGES: A two-month optimum growing season, heavy acid clay soil, and lots of rain. Frequent fruit-set failures in June because of cold nights, rain, and cool daytime temperatures.

TIPS: Start seeds early and grow seedlings to flowering stage under glass (or plastic), then move seedlings into containers to grow to maturity. Focus on early varieties. Grow plants in black 5-gallon containers (to absorb the heat of the sun). Mulch to keep soil off the lower foliage and avoid spreading soilborne diseases to plants.

Denise says that the general recommended time for transplanting in Vancouver is the Victoria Day weekend in late May; however, she's had good luck planting in containers at the beginning of May. Containers warm much faster than a garden and they often get a warm spell in May that gives the tomatoes a jump start.

BEST VARIETIES: Among her top performers and favorites for eating, Denise includes Eva Purple Ball, Paul Robeson, Aunt Ginny's Purple, Kimberly, and Galina. Tatiana has great success with Clear

Pink Early, Jaune Flamme, Black Krim, Kimberly, Black Cherry, Coyote, Cherokee Green, Pink Berkeley Tie-Dye, and Brandywine.

California

GARDENERS: Linda Black and Doug Frank

LOCATION: Fountain Valley (Orange County) and Fair Oaks (Sacramento area)

GROWING SEASON: With just a little precaution, Doug can be eating fresh — but not necessarily "vine-ripened" — tomatoes from mid-June through Thanksgiving. Some years, as in 2013, he's had them right up through December.

AVERAGE SEASONAL HIGH TEMPERATURES: 80s in Orange County and over 100°F in Sacramento

PEST/DISEASE PRESSURE: Doug battles nematodes in the garden (he minimizes their impact by working lots of organic material into his soil).

SPECIAL CHALLENGES: Linda deals with cool temperatures and coastal fog in the summertime, leading to fungal leaf problems. Doug's main challenge is dealing with often extreme temperatures. Every year, his region experiences temperatures exceeding 100°F. The plants don't appreciate the excess heat, new growth is burned, fruit is scorched, and production stops. This results in "blank spots" in the tomato season.

TIPS: Linda plants by the end of March to get plants established before the "June gloom" — the seasonal fog that causes foliage diseases to set in.

BEST VARIETIES: Among Doug's favorites are Gold Medal, Kosovo, Juliet, Sun Gold, and KBX. Linda has good luck with Dester, Joe's Pink Oxheart, Vorlon, Pink Honey, Chapman, Milka's Red Bulgarian, Red Barn, and Earl's Faux.

Texas

GARDENERS: Worth Doss and Michael Volk

LOCATION: Bastrop (Austin area) and El Paso (extreme western Texas)

GROWING SEASON: Typically from March to December. Michael has multiple growing seasons. He starts seed in December or January and plants out under row cover in early March for the first planting. For the second season, he starts seed of short-season varieties in July and plants out in late August or early September. If he had a greenhouse, he'd get a third season, starting seed of short-season varieties for planting out in late September.

Worth doesn't grow fall tomatoes because temperatures fall quickly in his area, leading to poor fruit set. Any fruit that does set risks freezing.

AVERAGE SEASONAL HIGH TEMPERATURES: In June and July, daytime high temperatures are in the 95°F range, easing off a bit to 90°F in August.

PEST/DISEASE PRESSURE: Michael's plants aren't affected by tomato diseases, but he does contend with aphids, webworms, and grasshoppers. He uses predators for pest control: green lacewings for the aphids and trichogramma wasps for webworm control. He also spreads diatomaceous earth around the containers for the ants that farm the aphids and NoLo bait, a grasshopper pathogen, around the yard every March.

Worth's plants can be afflicted by early and late blight, gray mold, and powdery mildew. He also deals with nematodes (in sandy soil), tomato hornworms, fruit-boring worms, stinkbugs, aphids, whiteflies, and mealy bugs, as well as small and large mammals such as opossums, raccoons, deer, squirrels, armadillos, and various birds.

SPECIAL CHALLENGES: Michael's main challenge is his very hot, very dry summer, with

temperatures sometimes exceeding 100°F in June, July, and August (and sometimes as early as May). He plants out early to avoid massive blossom drop. He suggests shading the plants for the summer grow-out; without it, his plants experience such tremendous blossom drop that he has little harvest. Michael's water has a high salt content, so he flushes his containers often with water processed through a reverse-osmosis filtering device.

TIPS: Focus on varieties with good foliage cover to minimize sunscald on the fruit. Mulch plants well and trim lower foliage to keep soilborne diseases from spreading. Keep food and water for wildlife in an area far away from the garden.

BEST VARIETIES: Because of a very brief window of optimum temperatures for fruit set, Worth tends to grow determinate and smaller-fruited varieties, some of which will set fruit all summer. He has particular success with cherry tomatoes (Matt's Wild Cherry and Sun Gold, for example), Black Plum, and Red and Yellow Pear. Michael succeeds with and enjoys Dwarf Jade Beauty, New Big Dwarf, Rosella Purple, Porter, and Glacier.

Florida

GARDENERS: Jan Matherly from Bradenton (central Gulf Coast) and Jessica Dubin of Boca Raton (southern Atlantic Coast)

GROWING SEASON: Jan starts her seeds in December so that she can harvest some tomatoes before the heat sets in for the summer. For a second harvest, she starts seeds in August, and the seedlings get off to a great start in the heat. Cooling nighttime temperatures in the fall allow for blooms that end up producing tomatoes for Thanksgiving and Christmas.

Jessica's main season stretches from late August/early September to late May/early June before disease pressure becomes overwhelming in the summer months. It's very rare for temperatures to drop near freezing during the winter.

AVERAGE SEASONAL HIGH TEMPERATURES: Low 90s in August and September

PEST/DISEASE PRESSURE: Jan and Jessica's main pests are aphids and silverleaf whitefly. Nematodes are sometimes a problem in Florida vegetable gardens as well (at least for gardeners planting in the ground, rather than in containers). Because the whitefly is a vector for tomato yellow leaf curl virus (TYLCV), Jessica uses fine-mesh insect netting over her plants to protect them from whitefly until the foliage growth warrants removal to allow aggressive pruning and increased airflow in mature plants. Sticky yellow traps and predator insects have failed in her garden due to overwhelming numbers of the pest insects and the prevalence of host plant species in the area.

SPECIAL CHALLENGES: Anyone who's familiar with this area of Florida knows about the constant humidity, which of course leads to gray leaf mold and various bacterial diseases. These are only enhanced by high temperature and wet foliage.

TIPS: Though spraying with an antifungal agent such as Daconil can hold off issues, it only seems to delay the inevitable. Good mulching, air circulation, and garden hygiene are all very important for maintaining plant health for as long as possible.

BEST VARIETIES: Jessica finds success with TYLCV-tolerant varieties like the commercially available Katana; they require less intensive maintenance and yield a more reliable crop than many other varieties in her garden. She also prefers heat-tolerant hybrids and heirlooms, finding consistent success with Kosovo, Amazon Chocolate, Carbon, Pink Berkeley Tie-Dye, Black Cherry, and Stupice. With extra care, she can grow even the most delicate and finicky varieties.

Eastern Canada

GARDENERS: Niki Jabbour and Neil Gillard

LOCATION: Halifax, Nova Scotia, and Holland Landing, Ontario (north of Toronto)

GROWING SEASON: 120 days for Niki, 100 days for Neil

AVERAGE SEASONAL HIGH TEMPERATURES: 70–80°F; the peak growing season high temperature is around 90°F.

PEST/DISEASE PRESSURE: Niki rarely has to deal with insect issues, and her main disease challenges are fungal.

Neil says that his plants generally experience diseases such as early blight, but he's never treated the plants and says they usually do well in spite of it. He rarely has pests, but in the past he has had to handpick hornworms and stinkbugs. A chance occurrence of whiteflies was treated to a blast of water from the hose.

SPECIAL CHALLENGES: Nova Scotia's short season and unpredictable weather — including summers that are frequently foggy, cool, and wet — can make growing tomatoes rather challenging. Niki grows mostly short-season varieties, with cherry types often providing the most reliable harvest.

TIPS: Niki starts seeds indoors in mid-March, moving the seedlings to the garden in late May. She pre-warms her garden soil with black plastic and enriches it with plenty of aged manure, kelp meal, and compost. Because the soil tends to be acidic, she also applies lime every autumn. To protect tomato seedlings and give them a jump start, Niki plants them in a plastic-covered mini hoop tunnel until the spring temperatures are settled and the risk of frost is well past. Once the covers are off and the tomatoes are growing well in the garden, she mulches the soil with a thick layer of straw or shredded leaves to regulate soil moisture and help reduce the occurrence of blight.

Because of his sandy soil, Neil has success growing tomatoes in containers. He starts seeds in early April and transplants later in the month, mulching his pots with wood chips to reduce the evaporation of water from the soil.

BEST VARIETIES: Some of Niki's favorite varieties include Sun Gold, Cherokee Purple, Costoluto Genovese, Big Rainbow, Persimmon, and Black Cherry. Neil enjoys Ashleigh, Casey's Pure Yellow, OTV Brandywine, Gary O'Sena, Golden Monarch, and Green Gage.

APPENDIX THE GREAT COMPETITION

Back in 1986, after several years of gardening, I finally grew tired of my hybrid-infused garden plot. Sure, some good hybrid tomatoes are available, ones that are more succulent than the types that most commercial growers seem to need to grow. But this did not satisfy my yearning for better-tasting, more interesting-looking vegetables.

The irony is that the answer was found in the tomatoes of the past. It was when I joined the Seed Savers Exchange and started to dabble in the multitude of treasures it made accessible that my gardening experience became one of fulfillment, wonder, and excitement. Needless to say, I jumped right in with total abandon. Ah, but what about all of those concerning warnings? Could I possibly grow these disease-prone, obsolete, low-yielding varieties successfully? Would I have anything to show for my efforts aside from some blemished, misshapen fruit on spindly, disease-ridden vines?

Being a scientist, I felt that the best way to approach this issue was to carry out an experiment. So, in 1987, I grew some of the most popular hybrid tomatoes side by side with a few of the more highly regarded and "famous" heirlooms. I kept detailed records of maturity dates, yields, fruit size, flavor, and observations about how each variety held up to disease. When all was said and done at the end of the growing season, I could compare the total number of tomatoes and total weight of fruit per plant and develop a feeling for the personalities of each variety with regard to flavor and visual interest. The results were so fascinating that my original one-year experiment with a limited number of varieties was expanded to three years and eventually involved lots of different tomatoes of all sizes, colors, and shapes. The best way to judge the results is to look at where my garden efforts have become focused, as I have now moved almost exclusively into growing open-pollinated varieties. There is simply very little, if any, reason for home gardeners to restrict themselves to hybrids.

The results of my three-year study are found in the tables on the following pages. If you had doubts about delving into the world of heirloom tomatoes, maybe this will persuade you to join in the fun. If you are already a convert, then this will just confirm what you already have discovered. These tables should also give you a good idea about what to expect from many tomatoes that you may be interested in but have never grown.

Most of the tomato varieties are indeterminate, grown vertically, and pruned to two or three main stems; the determinate (marked "D" in the tables) varieties were not pruned.

The idea was not to necessarily show that all heirlooms or open-pollinated tomatoes are superior in all respects to hybrids. In truth, the data show excellent and average examples in both categories. The data also show how variable the open-pollinated varieties are when compared to the somewhat more consistent (in terms of yield) hybrids. This, along with the fragile, relatively higher perishability, is probably why hybrids will always be more popular to commercial growers. Home gardeners do not have these concerns, however.

In 1987 and 1988, I was more concerned with comparison growing, as can be seen from the numbers of hybrids in the trials. By 1989, I had become convinced that open pollinated was the way to go, so I included fewer hybrids in the trials. To be fair, with the exception of Moreton, Supersteak, Early Cascade, Big Girl, and Ultra Sweet, the hybrids did very well in terms of yield and flavor. However, none of the hybrids were superior to the best of the open-pollinated varieties — Nepal, Brandywine, Anna Russian, and Polish, to name but a few of the superb heirlooms that I tested.

None of the hybrids came close to the yields from Ruby Gold, Yellow Bell, and Hugh's. On the flip side, none of the hybrids yielded as lightly as some of the heirlooms such as Ponderosa, Yellow

Brimmer, Rockingham, and Tice's Yellow Better Boy. The results indicated that the hybrids may be less fussy about the weather conditions of a particular gardening season, or the local climate.

In general, 1987 and 1989 seemed to be much better seasons for growing open-pollinated varieties than 1988, since the hybrids were more consistent that year in terms of yield. One possible problem was that Wayahead was diseased (it looked like tobacco mosaic virus), and the surrounding plants did not fare well.

Here are some generalizations from the data:

- In general, the open-pollinated tomatoes didn't show any more likelihood to become diseased as the season progressed. They bore fruit until killed by frost.
- By the end of the growing season, the condition of the foliage was comparable, even for the hybrids that were resistant to verticillium, fusarium, and nematodes.
- Some very good hybrid tomatoes are available, such as Better Girl (Northrup-King), Lemon Boy (widely available), Big Pick, Whopper, Better Boy (well known), and Gurney Girl. However, aside from the lovely canary-yellow Lemon Boy, hybrids have little diversity of color or flavor.
- In contrast, the variability within the open-pollinated tomatoes is immense, from the very bland (specifically bred for commercial growers) Fireball to the exquisitely rich flavor of Brandywine, and from the tiny Yellow Cherry to the huge Hugh's or Ruby Gold. And the colors are beyond belief.

Since 1989, I have grown hundreds of other open-pollinated tomatoes in my garden, and this year will continue the experiment. Even 25 years later, hybrid varieties play only a tiny part in my annual tomato garden. I have convinced myself that there is really no need to confine my gardening choices to the hybrid varieties, and perhaps I've convinced you as well!

Yellow White

APPENDIX THE GREAT COMPETITION

VARIETY	DAYS TO RIPEN	TOTAL NUMBER OF FRUITS	AVERAGE WT. OF FRUIT, OZ.	TOTAL WT. OF FRUIT, LBS	FLAVOR
1987 (Garden in Berwyn, Pennsylvania)					
Heirloom					
Yellow Cherry	63	773	0.2	9.25	A-
Tiger Tom	64	172	2	21.5	A-
Czech's Excellent Yellow	71	141	3	26.5	B+
Sugar Lump	72	581	0.5	18.1	B+
Fireball (D)	68	58	5.6	20.6	B-
Veeroma (D)	80	220	2.5	34	B-
Super Marmade (D)	79	81	5.5	28	B+
Nepal	82	60	7	26.5	A
Pineapple	85	23	14.6	22.3	B+
Persimmon	86	39	13.5	33	B+
Ruby Gold	83	30	16.6	31.3	B+
Brandywine	77	32	8.4	16.7	A-
Abraham Lincoln	63	69	5.2	22.2	B+
Hybrids					
Better Girl	65	49	6.5	19.7	A
Moreton	80	40	5.6	14	B
Lemon Boy	89	58	6.9	25.2	A
Ultra Boy	83	35	9.7	21.3	A-
Supersteak	85	13	13.5	11	B+
1988 (Garden in Berwyn, Pennsylvania)					
Heirloom					
Hungarian Ital. (D)	78	24	3.5	5.1	B-
Bisignano #2	67	45	8.2	23.1	A-
Glesener	85	25	9.9	15.6	A-
Brandywine	89	16	10.8	10.8	A-
Abraham Lincoln	67	57	6	18.7	B+
Sabre	82	34	9.9	21.1	B
Ponderosa	80	22	8.1	11.2	B

(D) = determinate

VARIETY	DAYS TO RIPEN	TOTAL NUMBER OF FRUITS	AVERAGE WT. OF FRUIT, OZ.	TOTAL WT. OF FRUIT, LBS	FLAVOR
Yellow Brimmer	101	8	16	8	B+
Valencia	74	33	8.2	16.9	B+
Peron	74	17	7.5	7.9	A-
Oregon Spring (D)	70	63	3.2	12.8	B
Rutgers	87	33	9.8	20.1	B+
Wayahead (D)	66	35	2.5	5.6	B-
Hybrids					
Big Pick	72	44	6	16.4	A
Early Cascade	66	90	3.3	18.8	B-
Whopper	73	49	6.2	18.9	A-
Better Boy	67	47	7.1	20.8	A
Firebird	68	50	6	18.8	A-
Gurney Girl	67	34	6.5	13.9	A
Big Girl	73	57	6.6	23.5	B
Ultra Sweet	67	44	6.7	18.1	B
1989 (Garden in West Chester, Pennsylvania)					
Heirloom					
Ester Hess	67	739	0.6	27.7	B
Yellow Bell	70	245	3	45.9	A-
Hunt Family Favorite	57	57	8.1	28.9	A
Fritsche	71	56	5.8	20.3	A-
Lillian's Red Kansas Paste	70	45	7.1	19.5	A
Sutton	58	44	8.1	22.4	B+
Banana Legs (D)	59	43	3	8.1	B
Viva	78	39	10.9	26.5	B+
Hugh's	81	35	19.5	42.7	A-
Holy Land	71	32	10.3	20.5	B
Anna Russian	68	34	8.4	17.9	A
Rockingham	67	28	6.5	11.3	A-

(D) = determinate

THE GREAT COMPETITION

VARIETY	DAYS TO RIPEN	TOTAL NUMBER OF FRUITS	AVERAGE WT. OF FRUIT, OZ.	TOTAL WT. OF FRUIT, LBS	FLAVOR
Firesteel	73	34	7.2	15.4	B+
Wolford Wonder	67	28	15.9	27.8	B
Polish	73	23	15.5	22.3	A
German Garden Time	80	23	14.6	20.9	B-
Believe It or Not	71	21	19.2	25.3	A-
Georgia Streak	68	21	16.1	21.1	B+
Goldie	70	27	14.9	25.2	B+
Golden Oxheart	77	27	8.9	15	A
Tappy's Finest	72	23	16	22.9	A-
Old Brooks	69	19	10.7	12.7	A-
Pruden's Purple	72	18	13.6	15.3	A-
Mortgage Lifter, Pesta Strain	75	17	22.1	23.4	B+
Tice's Yellow Better Boy	67	13	9.4	7.6	B+
Yellow Oxheart	79	16	10.9	10.9	A
Lillian's Yellow Heirloom	103	11	16.7	11.5	A-
Yellow White	70	13	17.6	14.3	B+
Andrew Rahart's Jumbo Red	87	10	12	7.5	A-
Hybrids					
Sweet Million	59	1,045	0.5	32.5	A
JSS361 (D)	62	22	6	8.3	A-
Jumbo Tom	80	15	19.5	18.3	A-
Valley Girl (D)	74	52	6.7	21.8	B

(D) = determinate

RESOURCES AND SOURCES

Recommended Reading

Books

Burr, Fearing, Jr. *The Field and Garden Vegetables of America*. Crosby and Nichols, 1863.

DuBose, Fred. *The Total Tomato*. Harper Collins, 1985.

Goldman, Amy. *The Heirloom Tomato: From Garden to Table*. Bloomsbury, 2008.

Hendrickson, Robert. *The Great American Tomato Book*. Doubleday, 1977.

Jabbour, Niki. *The Year-Round Vegetable Gardener*. Storey Publishing, 2011.

Livingston, A. W. *Livingston and the Tomato*. A. W. Livingston's Sons, 1893.

Male, Carolyn J. *100 Tomatoes for the American Garden*. Workman, 1999.

Ortho Book Divions. *All About Tomatoes*, rev. ed. Chevron Chemical Company, 1977.

Smith, Andrew F. *The Tomato in America*. University of South Carolina Press, 1994.

Vilmorin-Andrieux. *The Vegetable Garden*, 3rd ed. E. P. Dutton, 1920.

Whealy, Diane Ott. *Gathering: Memoir of a Seed Saver*. Seed Savers Exchange, 2011.

Whealy, Kent, and Arllys Adelmann, eds. *Seed Savers Exchange: The First Ten Years, 1975–1985*. Seed Savers Exchange Publications, 1986.

Literature and Bulletins

Bailey, L. H. "Notes on Tomatoes." Agricultural College of Michigan, Bulletin 19, 1886, and Bulletin 31, 1887.

Bailey, L. H., and L. C. Corbett. "Tomatoes," Bulletin 45. Cornell University Agricultural Experimental Station, 1892.

Ballard, W. R., T. H. White, and C. P. Close. "Miscellaneous Greenhouse Notes," Maryland Agriculture Experimental Station, Bulletin 127, 1908.

Morrison, Gordon. "Tomato Varieties." Michigan State College Agricultural Experiment Station, 1938.

Myers, C. E. "The Effect of Selection in the Tomato," Bulletin 248. Pennsylvania Agricultural Experimental Station, 1930.

_____. "Strain Tests of Tomatoes," Bulletin 129. Pennsylvania State College Agricultural Experimental Station, March 1914.

_____. "A Variety Test of Tomatoes." Pennsylvania State College, Ag. Experimental Station, Separate 21 from the Annual Report for 1913–1914, published 1916.

Sapers, G. M., J. G. Phillips, and A. K. Stoner. "Tomato Acidity and the Safety of Home Canned Tomatoes." *HortScience* 12, no. 3. (June 1977): 204–208.

Taft, L. R. "Vegetables: Tests of Varieties and Methods of Culture." Michigan Agricultural Experiment Station, Bulletin 70, 1891.

Recommended Seed Companies

Baker Creek Heirloom Seed Co.
417-924-8917
www.rareseeds.com

Casey's Heirloom Tomatoes of Airdrie
jwlcasey@yahoo.ca
www.caseysheirloomtomatoes.ca

Fedco Seeds
207-426-9900
www.fedcoseeds.com

Heirloom Seeds
724-663-5356
www.heirloomseeds.com

Heritage Tomato Seed
http://heritagetomatoseed.com

Johnny's Selected Seeds
877-564-6697
www.johnnyseeds.com

J.W. Jung
800-297-3123
www.jungseed.com

Laurel's Heirloom Tomato Plants
310-534-8611
www.heirloomtomatoplants.com

Peaceful Valley Farm Supply
888-784-1722
www.groworganic.com
Grow bags for container gardening.

Pinetree Garden Seeds
207-926-3400
www.superseeds.com

The Sample Seed Shop
716-871-1137
www.sampleseeds.com

Sand Hill Preservation Center
563-246-2299
www.sandhillpreservation.com

Seed Savers Exchange
563-382-5990
www.seedsavers.org

Seeds 'n Such
803-663-1501
www.seedsnsuch.com

Selected Plants.com
205-921-4180
www.selectedplants.com

Southern Exposure Seed Exchange
540-894-9480
www.southernexposure.com

Stokes Seeds
800-396-9238
www.stokeseeds.com

Tatiana's TOMATObase
tatianak.garden@gmail.com
http://tatianastomatobase.com

Territorial Seed Company
800-626-0866
www.territorialseed.com

Tomato Growers Supply Company
888-478-7333
www.tomatogrowers.com

Totally Tomatoes
800-345-5977
www.totallytomato.com

Tomatofest
www.tomatofest.com

Victory Seed Company
503-829-3126
www.victoryseeds.com

White Harvest Seed Company
866-424-3185
http://whiteharvestseed.com

Useful Websites and Blogs

Nctomatoman's Dense Seed Planting Technique
YouTube
www.youtube.com/watch?v=CoYgX3y5ptQ

Tatiana's TOMATObase
http://tatianastomatobase.com

Tomatoville
www.tomatoville.com
Discussion board

Vegetable MD Online
Department of Plant Pathology, Cornell University
http://vegetablemdonline.ppath.cornell.edu

***The Year Round Vegetable Gardener* blog**
http://yearroundveggiegardener.blogspot.com

Information on Raising/Lowering pH Levels

Applying Lime to Raise Soil pH for Crop Production (Western Oregon)
http://ir.library.oregonstate.edu/xmlui/bitstream/handle/1957/38531/em9057.pdf
Written by N. P. Anderson, J. M. Hart, D. M. Sullivan, N. W. Christensen, D. A. Horneck, and G. J. Pirelli. Bulletin EM 9057. Oregon State University Extension, May 2013.

Changing pH in Soil
University of California Cooperative Extension
http://vric.ucdavis.edu/pdf/Soil/ChangingpHinSoil.pdf

Lower Soil pH
Soil Testing Laboratories, University of Wisconsin, Madison
http://uwlab.soils.wisc.edu/pubs/lowering_ph.pdf

Lowering Soil pH for Horticulture Crops
www.extension.purdue.edu/extmedia/HO/HO-241-W.pdf
Written by Michael V. Mickelbart and Kelly M. Stanton. Bulletin HO-241-W. Purdue University Cooperative Extension Service, April 2012.

GLOSSARY

F_1, HYBRID, OR F_1 HYBRID. A hybrid tomato (synonymous with F_1, meaning first filial generation) variety is created by crossing two other varieties. Pollen from one parent (the male) is applied to the style of the other parent (the female). The tomato that develops contains the F_1 seeds.

OPEN POLLINATED. A term used to describe varieties that are genetically uniform and that reproduce the parent from saved seeds. It originated in the 1920s to describe the pollination of a flower without human interaction.

HEIRLOOM. This is a particular category of open-pollinated varieties that pre-date the majority of modern hybrid varieties. There is no definite age qualification for a vegetable to be considered an heirloom. For tomatoes, I choose 1949 as the key date, because it marks the introduction of Big Boy.

SUCKER OR SIDE SHOOT. This is the growth that emerges at a 45-degree angle between the main stem and leaf stem.

COTYLEDON LEAVES. Also known as seed leaves, these are the first two (or, rarely, three) smooth-edged leaves that emerge upon germination of tomato seeds

TRUE LEAVES. These are the leaves that emerge after the cotyledon leaves; they are often distinctly shaped for particular tomato varieties.

INDETERMINATE. Tomato varieties that grow indefinitely until killed by frost or disease are called indeterminate. The main stem and side shoots can easily extend to 10 feet or more by the end of the growing season.

DETERMINATE. Determinate tomato varieties have compact growth of 2 to 4 feet in width and height. Growth stops when flowers set at the end of the main stem and side shoots. Because of this, determinate plants should not be pruned, in order to achieve optimum yield.

DWARF. Dwarf tomato varieties have thick central stems, dark bluish-green crinkly foliage, and slow vertical growth. In fruit habit, they behave most like very slowly growing indeterminates, reaching 3 to 5 feet by the end of the growing season, depending upon the variety.

REGULAR LEAF. Tomato varieties whose leaves are toothed or serrated at the edge are called regular-leaf varieties.

POTATO LEAF. If the edges of the leaves of a tomato variety are smooth, rather than serrated, it is known as a potato-leaf variety, because of its similarity to potato plant foliage.

RUGOSE FOLIAGE. Leaves that are somewhat crinkled looking. This is a trait of dwarf tomatoes, in particular.

OBLATE. Tomatoes that are wider than they are deep (the distance from stem attachment to blossom end) are often called oblate.

SEGREGATION. Segregation is the appearance of distinct genetic traits from each parent, as they occur in the individual offspring of F_1 hybrids. For example, if the F_1 hybrid was the result of a cross of a regular-leaf variety with a potato-leaf variety, some of the offspring seedlings will exhibit potato-leaf foliage, and others will have regular-leaf foliage.

SELECTION. When a breeder is working with a planting of tomatoes, selection is the process of choosing one particular plant to continue breeding efforts with, because it exhibits a different characteristic (often an advantage over the rest of the population).

METRIC CONVERSION CHART

Converting Recipe Measurements to Metric

Use the following formulas for converting US measurements to metric.

Weights/Measures

When the Measurement Given Is	Multiply It By	To Convert To
teaspoons	4.93	milliliters
tablespoons	14.79	milliliters
fluid ounces	29.57	milliliters
cups	236.59	milliliters
cups	0.236	liters
pints	473.18	milliliters
pints	0.473	liters
quarts	946.36	milliliters
quarts	0.946	liters
gallons	3.785	liters
ounces	28.35	grams
pounds	0.454	kilograms

Lengths

When the Measurement Given Is	Multiply It By	To Convert To
inches	2.54	centimeters (cm)
inches	25.4	millimeters (mm)
inches	0.0254	meters (m)
feet	0.305	meters (m)
yards	0.9144	meters (m)
yards	91.44	centimeters (cm)

ACKNOWLEDGMENTS

This book is the culmination of many years of gardening, with countless tomatoes grown and tasted. The one constant throughout all of those years has been my wife, Susan. The endless support, prodding, and, especially, patience that she provided, starting with our very first garden just after our wedding, made this book, as well as the deep happiness that I experience daily, possible. Susan knew I had this book in me for many years, and, finally, here it is. Susan, my love for you only grows each day, and I couldn't have done it without your faith in me.

I was fortunate to have expert tomato tasters, and occasional garden helpers, as daughters. Instead of thinking of their dad as "unusual," they not only took part in and enjoyed our annual tomato bounty but developed a love of gardening themselves. Sara and Caitlin, I am absolutely delighted with who you are, and how you are living your lives.

My love of gardening was initiated by my dad, Wilfred — a noted heirloom tomato grower himself in the twilight of his life — and grandfather, Walter. In fact, my gramps's tomatoes were the ones that converted me to a lover of the fruit (or is it a vegetable?). His were the first I dared to taste, and it was love at first bite.

My tomato passion was stimulated, molded, and energized by many people over the years. Certainly, the interest expressed by Diane Ott Whealy and Kent Whealy during my early years as a member of the Seed Savers Exchange went a long way in giving me the confidence to jump in with both feet. Of my various gardening friends through the years, none has been as steadfast in his support and interest as Jeff Fleming. Many years after meeting at Dartmouth, Jeff and I can still spend hours discussing tomatoes, music, and life in general. Also crucial to my early tomato years was Carolyn Male, who spends countless hours providing useful advice to tomato growers everywhere.

It was always astounding to me that my out-of-control hobby was of such interest to so many, but I did find myself walking through my garden with many people along the way, the result being lovely articles in local newspapers and magazines. I deeply appreciate the kind treatment received from Jane Pepper, John Snow, Jack Ruttle, Debbie Moose, A. C. Snow, and Pam Beck in those early, formative years of my obsession with tomatoes. Many others followed and treated me very well in print, but you were the first to spread the word about what I do.

Each spring since the late 1990s, I get to reunite with so many of our "annual tomato friends" — our vegetable seedlings customers — at the Raleigh Farmers Market or in our driveway. Spreading the stories and joy of growing heirloom vegetables to them has been a joy and helps to energize my own gardening efforts. My wife, Susan, and I have watched families and hobbies grow before our very eyes, and we feel honored and fortunate to have been able to do so.

One of the most interesting recent pieces of my tomato puzzle is the Dwarf Tomato Breeding Project, and my project co-leader, Patrina Nuske Small. The project would not have thrived without a way to discuss it collaboratively with all of our many participants, and the Tomatoville website created and managed by Craig Shea is a critical success factor. To the hundreds of volunteers all over the world, thanks for the hard work and exciting results.

Tomatopalooza provided a way to share tomato-tasting experiences locally. Lee Newman, the event cofounder, Lori Wheatley and Brian Sadler, whose organization and

photography of many of the events led directly to their success, and Jimmy Holcomb and Fred Stewart, hosts of the most successful examples, all feed into the experiences that are this book.

I've met so many others who have contributed to my tomato hobby/addiction/profession. In no particular order, among them are Bill Minkey, Neil Lockhart, Ira Wallace, Sarig Agassi, Jeff and Patti McCormack, Rob Johnston and Janika Eckert, Mike and Denise Dunton, Tatiana Koucharevna, Ted Maiden, Darrel Jones, Keith Mueller, Alex and Betsy Hitt, and all of our wonderful tomato seedling customers and various hosts and hostesses to tomato lectures. Special thanks go to Nancy Butterfield and Reitzel Deaton for answering the tomato SOS for this book. For a hobby that is in the midst of a 35-year run, so many helped me along the way, and for anyone I've left out by name, please understand how much I appreciate you.

My two best Raleigh-area friends, Allen Jones and Bob Willoughby, exhibited great patience and attention during our weekly lunch sessions over many years as I regaled them with various tomato stories. I deeply appreciate their encouragement, support, and enduring friendship.

Finally, thanks go to Niki Jabbour for tapping my tomato brain for occasional spots on her wonderful radio show and making the connection with Storey Publishing; to Stephen Garrett, Kip Dawkins, and Marcie Blough for the incredible photographs; to Carleen Madigan for superb editing support and being so gentle and patient with her requests to me; and to Carolyn Eckert for her art direction of this book.

LETTERS FROM SEED SAVERS

I consider myself fortunate indeed to have received valuable family heirloom seeds from so many gardeners through the years. I've kept all of the letters and consider them priceless. This small sampling shows how unique — and rich with stories — they are.

Mexico Midget

I'm sending some [seeds] of probably the world's smallest tomato. My brother in New Mexico, who is a truck driver to Texas (with hay) got these seeds there, said he was told that they came from Mexico. They are more of a josh tomato with a real sweet flavor but is mostly juice & seeds. They don't mix with any other tomato and come up year after year. . . .They might be wild. They don't run out until frost. Nothing hinders them.

—Barney Laman, Chico, CA

Cherokee Purple

I talked with the lady that gave me the purple tomato seed. She got them from her neighbors several years ago. The only thing they know about them is that the tomato has been in their family for about 100 years and they were gotten from the Cherokee Indians. . . . I'm sorry I can't tell you anything more about them.

— J.D. Green, Sevierville, TN

Mr. Le Houllier,
I talked with the lady that gave me the purple tomato seed. She got them from her neighbors several years ago. The only thing they know about them is that the tomato has been in their family about 100 yrs and they were gotten from the Cherokee ~~Indian~~ Indians.
I sure am glad you like them, mine aren't ripe yet, going to be late this year. Sure hope frost doesn't get them. I sent Glenn Drowns some too.
I have a couple of others I am going to send you, but they will not be anything like the purple one, I can't believe I have something new that will be offered thru the SSE.
I'm sorry I can't tell you anything more about them.

J.D. Green

JAMES R. HALLADAY
LINDA S. HALLADAY
5994 STERRETTANIA RD.
FAIRVIEW, PA 16415

Dear Dr. LeHoullier,

I am enclosing the requested tomato seeds for Pineapple and Mortgage Lifter as well as seed from Tiger Tom and Czech's excellent yellow.

Mortgage Lifter: developed during 1920's or 1930's. This strain was obtained from the Ashland, Ky/Huntington, West. Va. area where it has been raised for several generations. It is a large pink skinned tomato resembling a beefsteak type with a sweet mild flavor and relatively few seeds. It has a long maturation period and should be started indoors. It is somewhat slow to develop but is quite vigorous once it gets going.

Pineapple: History unknown. Seeds were given to my Father-in-Law by a neighboring farmer near Elizabethtown, Ky. This is an extremely large yellow tomato with a pink area around the blossom end and pink marbling throughout the middle (inside). It's growth characteristics are similar to Mortgage lifter. This is not listed in any seed catalogues and has been maintained in the locality in Ky where I got it from for generations as nearly as I can tell.

Tiger Tom: This and Czech's excellent Yellow both originated with a czechloslovakian tomato breeder and would have died with him if it had not become part of the Ben Quisenberry collection and then from there, a part of the SSE collection. Both are golf ball size or larger and early to mature. Both are extremely prolific and productive for me here in Pa. Tiger Tom is red skinned with yellow stripes and is not only unusual and attractive, but is very rich in "Tomato flavor." I start it indoors about March 15 and usually have tomatoes ripe by July 15–20.

Mortgage Lifter

Mortgage Lifter: developed during 1920s or 1930s. This strain was obtained from the Ashland, KY/Huntington, WV area, where it has been raised for several generations. It is a large, pink-skinned tomato resembling a beefsteak type with a sweet mild flavor and relatively few seeds. It has a long maturation period and should be started indoors. It is somewhat slow to develop, but is quite vigorous once it gets going.

— James Halladay, Fairview, PA

1. Charlotte Mullens
Rt. 1 Box 304K
Summersville, WV
26651

Dear Craig,

Thank you for the seeds. I have enclosed the ones you wanted to try. And also my morgage lifter it is a large Dark pink and very delicious. They don't always grow smooth but when you see their size you get Surprised. My Dad is 84 and mom is 74 and for family use they won't grow anyother kind. They raise smooth ones to take to the farmers Market but keep all the Morgage lifters for themselves. I told them about all my different kinds of seeds and Dad said "Why?" I guess a person can get their mind set and thats that.

Maybe someday when you get around to it you can send me a few of the following ones on you list. Don't be in a hurry I'm going to be busy trying out the new ones I've Collected this year so far. I'll have little Stakes every where. (Ha Ha)

Brandywine	Tappy's Finest
Hunt family Favorite	Believe it or not
Old Brooks	Amish Paste
Dinner Plate	Kentucky Heirloom

Thank you for the seeds. I have enclosed the ones you wanted to try and also my Mortgage Lifter. It is a large dark pink and very delicious. They don't always grow smooth but when you see their size you get surprised. My dad is 84 and mom is 74 and for family use, they won't grow any other kind. They raise smooth ones to take to the farmers market but keep all the Mortgage Lifters for themselves. I told them about all my different kinds of seeds and Dad said "Why?" I guess a person can get their mind set and that's that.

— Charlotte Mullens, Summersville, WV

Anna Russian Tomato

Dear Sir

I have enclosed a small sample of tomato seeds. This variety was given to my grandfather (Kenneth Wilcox) a number of years ago by a Russian immigrant whose family sent him the seed. I have been growing it for several years now but our growing season is not quite long enough for the majority of the fruit to ripen.

I haven't kept very good notes, but this is my recollection of the fruit & plant.

OR HI B
Brenda Hillenius
505 NW 11th #1
Corvallis, OR 97330

- indeterminate / large vine
- fruit: heart shaped, large - 3-5" diameter, pinkish in color, low acid(?), meaty, thin skinned, prone to cracking in rain or high moisture.
- long growing season 110 days? from transplant

please grow these seeds & send me a note on how they do for you. I'm hoping to offer this variety through SSE next year. This year I have such a small quantity of seed I'm only sending it to [illegible] people. THANKS! -Brenda

Anna Russian

I have enclosed a small sample of tomato seeds. This variety was given to my grandfather (Kenneth Wilcox) a number of years ago by a Russian immigrant whose family sent him the seed. I have been growing it for several years now but our growing season is not quite long enough for the majority of the fruit to ripen. . . . Please grow these seeds and send me a note on how they do for you.

—Brenda Hillenius, Corvallis, OR

INDEX

Page numbers in *italic* indicate photos or illustrations; page numbers in **bold** indicate charts.

D

INDEX

E

F

INDEX

J

K

L

Q

R

S

INDEX

INTERIOR PHOTO CREDITS

© Bridgeman Photo Library, 22, *Still Life with Vegetables*, 1826 (oil on canvas), Peale, James the Elder (1749–1831)/Museum of Fine Arts, Houston Texas, USA/The Bayou Bend Collection, museum purchase funded by the Theta Charity Antiques Show in honor of Mrs. Fred T. Couper, Jr.

© GAP Photos, Ltd./Jo Whitworth, 205 (top)

© GAP Photos, Ltd./Robert Mabic, 222 (top)

© Garden World Images/www.agefotostock.com, 215 (right)

© Jim Corwin/Alamy, 196 (right)

© John M. Coffman/Science Source, 196 (left)

© MAP/Nathalie Pasquet/www.agefotostock.com, 207 (left)

Mars Vilaubi, 4, 13, 26, 27, 35 (top left), 38, 39, 50–53, 69, 101, 110, 111, 115 (middle right), 119, 122 (top), 127, 146, 147, 152–155, 177, 236, 237, 247–255

© Martin Hughes-Jones/Alamy, 222 (bottom)

© Sean Russell/Getty Images, 176

© Shoe Heel Factory, inside front, 1–3, 6–8, 10, 16–18, 20, 21, 28–31, 33, 35 (except top left), 36, 40, 44, 45, 47–49, 54–55, 58, 62–63, 66, 71 (bottom), 83, 87, 93, 94, 98–100, 103 (right), 104–105, 108, 109, 113, 115 (except top, middle, and bottom right), 117, 118, 121, 124–125, 128, 131, 132, 138–143, 149, 150, 156–158, 169, 180 (bottom), 181, 182, 189, 190, 192, 198–199, 202, 207 (right), 208 (bottom), 210, 211 (right), 213, 214, 216–219, 221 (top), 239, 245

© Stephen L. Garrett, 5, 46, 56, 59–61, 67, 68, 70, 71 (top), 72–76, 78, 80–82, 85, 86, 88–90, 92, 95, 102, 103 (left), 115 (top and bottom right), 116, 122 (except top), 123, 134–137, 151, 159, 160, 162, 164–166, 168, 170-171, 180 (top), 181, 189, 191, 195, 201, 203, 205 (bottom), 208 (top), 211 (left), 212, 221 (bottom), 229, inside back

© Tim Gainey/Alamy, 79

© Trevor Sims/www.agefotostock.com, 215 (left)

MORE GREAT VEGETABLE GARDENING BOOKS *from Storey*

CARROTS LOVE TOMATOES *by Louise Riotte*
Plant parsley and asparagus together, and you'll have more of each. But keep broccoli and tomato plants far apart if you want them to thrive. This classic companion-gardening guide shows you how to use the native attributes of vegetables to create a naturally bountiful garden. Learn which plants nourish the soil, which repel pests, which encourage each other, and which just don't get along.
224 pages. Paper. ISBN 978-1-58017-027-7.

TOMATO *by Lawrence Davis-Hollander*
Now that you're growing some epic tomatoes, it's time to discover new ways to enjoy them with this delicious collection of 150 tempting recipes for appetizers, preserves, soups, salads, entrees, and even desserts. Celebrity chef contributors include Alice Waters, Deborah Madison, Daniel Boulud, Rick Bayless, Melissa Kelly, and many others.
288 pages. Paper. ISBN 978-1-60342-478-3.

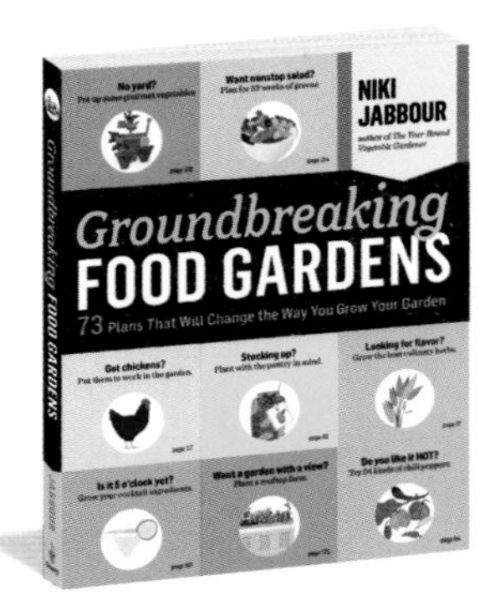

GROUNDBREAKING FOOD GARDENS
by Niki Jabbour
Vegetable gardens can be designed for flavor and fun! Niki Jabbour, author of the best-selling *The Year-Round Vegetable Gardener,* has collected 73 plans for novel and inspiring food gardens from her favorite superstar gardeners, including Amy Stewart, Amanda Thomsen, Barbara Pleasant, Dave DeWitt, and Jessi Bloom. Each plan is fully illustrated and includes a profile of the contributor, the story behind the design, and a plant list.
272 pages. Paper. ISBN 978-1-61212-061-4.

THE WILDLIFE FRIENDLY VEGETABLE GARDENER *by Tammi Hartung*
This one-of-a-kind book shows you how to create a peaceful co-existence between your vegetable garden and the wildlife who consider it part of their habitat. By understanding and working with the surrounding environment — instead of continually fighting it — you'll reap a larger harvest with less stress and effort.
144 pages. Paper. ISBN 978-1-61212-055-3.

These and other books from Storey Publishing are available wherever quality books are sold or by calling 1-800-441-5700. Visit us at www.storey.com or sign up for our newsletter at *www.storey.com/signup*.